D0857699

The Greek cosmologists

To
Phyllis

The Greek cosmologists

Volume 1

The formation of the atomic theory and its earliest critics

David Furley
Charles Ewing Professor of Greek Language and Literature, Princeton University

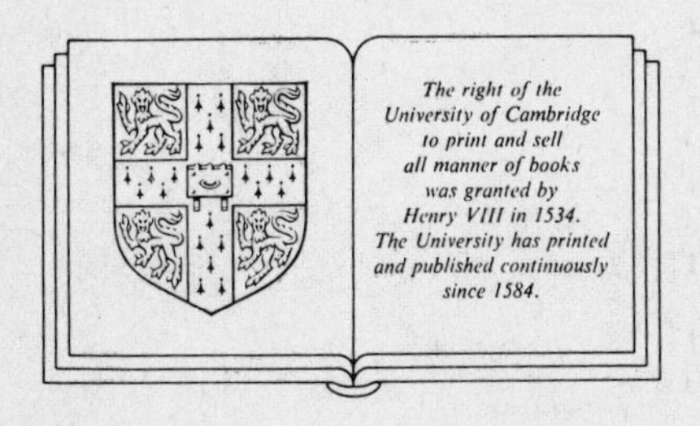

Cambridge University Press
Cambridge
London New York New Rochelle
Melbourne Sydney

Published by the Press Syndicate of the University of Cambridge
The Pitt Building, Trumpington Street, Cambridge CB2 1RP
32 East 57th Street, New York, NY 10022, USA
10 Stamford Road, Oakleigh, Melbourne 3166, Australia

First published 1987

Printed in Great Britain by
Redwood Burn Limited, Trowbridge, Wiltshire

British Library cataloguing in publication data

Furley, David
The Greek cosmologists.
Vol. 1: The formation of the atomic
theory and its earliest critics
1. Cosmology, Greek
I. Title
113'.0938 B187.C7

Library of Congress cataloguing in publication data

Furley, David J.
The Greek cosmologists.
Bibliography.
Includes index.
Contents: v. 1. The formation of the atomic
theory and its earliest critics.
1. Cosmology, Ancient. I. Title.
BD495.F87 1987 113'.0938 86-26384

ISBN 0 521 33328 8

RB

Contents

Preface

This volume tells one side of the story of ancient Greek cosmology, and takes it as far as the criticisms directed by Aristotle at his materialistic predecessors. The book is to be completed in a second volume, *The Teleological World Picture and its Opponents*. In that I shall give an account of the cosmology of Plato's *Timaeus* and Aristotle's works on the philosophy of nature, followed by the attempted rehabilitation of the atomic theory of the universe by Epicurus and his followers. Then I shall try to describe the arguments of philosophers of the Hellenistic period and later, including the Stoics and the later Peripatetics. I hope to take the story as far as the time of Simplicius and Philoponus, in the sixth century A.D.

I have tried to make the book readable by those with no specialized knowledge of Greek philosophy, and no Greek. Although it is not designed to be a course textbook, in writing it I have kept in mind the students to whom I lecture every year in Princeton University in a course called 'Introduction to Ancient Philosophy'. They include some who are taking a major in Classics or Philosophy, but also many who have never before studied the ancient world or read anything of the Greek philosophers. I have tried to keep the philological technicalities to a minimum.

On some subjects I have adopted positions that need more detailed defense than I have been able to give in this context. I have sometimes argued the case in articles that have been published separately in the professional journals, or in papers that are not yet published. I hope to collect these together in one volume, and to publish them under the title *Cosmic Problems*. References are given in the footnotes to those already published.

Although I have tried to write a book accessible to anyone who is interested in the history of science and philosophy, I hope there may be something in it that will interest my professional colleagues too. Much of the territory covered is certainly quite familiar. But the particular selection of subject matter, and the thesis that one can profitably treat Greek cosmology as an opposition between two fundamentally opposed ways of looking at the world, make this book a little different. It is very nearly seventy-five years since Pierre Duhem launched his great *Système du monde* with a volume on Greek (mainly Platonic and Aristotelian) cosmology; there has been much research since then, especially on the period before Plato, as well as a vast change in our own cosmological outlook. So it may well be time for another attempt to take a comprehensive view of the subject.

I must acknowledge, with much gratitude, the help of many friends too numerous to mention here by name. Those who study classical philosophy form a relatively small enclave in the world of learning. It is a friendly place to live in, and I have enjoyed and profited from personal contacts with many of those whose names appear in the Bibliography at the end of this volume, as well as from their writings. Those who have had a special connection with this book are Tony Long, who read an earlier version in typescript, and Richard Sorabji, who read a good deal of the present version. Both made valuable suggestions for which I am very grateful. I also thank an anonymous reader for the Cambridge University Press, who made me think harder and (I hope) better about a number of points.

Three of my graduate students, future colleagues, have given me great help in the preparation of the typescript. Jeffrey Purinton interpreted the handwritten draft of several chapters and put them on the computer. Carolyn Higbie typed other chapters and made the first draft of the Bibliography and Index of Passages. Paul Vander Waerdt read the whole typescript and helped to remove many errors of commission and omission.

I must also acknowledge the help of several institutions. The early stages of my work on the theme of matter and motion in classical antiquity were supported by a grant from the National Science Foundation, which enabled me to supplement a sabbatical leave from Princeton University and spend the year 1973/4 free of teaching responsibilities except for a graduate seminar on the subject. In 1978/9 I received a Fellowship from the National Endowment for the Humanities, and also had the stimulus and pleasure of teaching a Summer Seminar for College Teachers funded by the Education Division of N.E.H. The manuscript of this volume was completed in the spring of 1984, in another semester of sabbatical leave. I am grateful to all these institutions, and especially to Princeton University for the unfailing support I have received from librarians and administrators, students and colleagues.

D.J.F.

Princeton
1986

1 Two pictures of the world

The educational tradition of the Western half of the world has fairly recently adopted a new picture of the physical universe and man's place in it. We are now brought up to believe that we live on one of the planets that move in orbits around a rather minor star, itself a member of one galaxy among a vast number of others. We are taught that our planet is not eternal, but had an origin and will have an end; that it has grown into its present condition through a long evolutionary process, in which the minerals developed first, before any life appeared, and then the complex living species that we see around us, including man, evolved over a very long period of time through the processes of mutation, natural selection, and inheritance.

This picture has largely, though not entirely, replaced an earlier one that held the field for many centuries. According to that, the earth is stationary at the center of the cosmos. The stars, planets, sun, and moon revolve around the earth. The cosmos is finite, being bounded by the outermost sphere of the heavens, and is unique. This cosmic order, according to the prevalent, Judaeo-Christian version of the picture, was determined and brought into being at a certain point of time by God, who adopted it because it is a good order. He also created all the living species that now exist and arranged them in a hierarchical order of decreasing complexity or excellence, with man at the top of the scale. 'Above' the whole order of earthly creatures is a rather similar order of heavenly beings, all better than man but having similar characteristics, with God at the top. So the species Man, being created in the image of God, and being the best of the creatures that live near the center of the whole universe, has a uniquely privileged position.

Elements of this older picture of the world still survive in religious education and in popular thought, speech and action. On the whole, it has been banished from research laboratories and departments of philosophy in universities, although there are some highly sophisticated attempts to reconcile the new world picture with certain features of the old, especially the creative role of God and the special relationship between God and man.

The transition from the older to the newer picture was, of course, not instantaneous or even rapid; it occupied several centuries, beginning (although to settle upon any single beginning or end is rather arbitrary) with Copernicus' proposal to remove the earth from the center of the universe, continuing with Darwin's theory of the evolution of living species, and culminating in the new scientific cosmology of the 1980s – the so-called 'Big Bang' theory of the origin of the universe.

What I have called the 'older' world picture was worked out in essence in the pre-Christian era: it is largely the work of Aristotle, although there are, naturally enough, important differences between the Aristotelian and Christian versions. But it happens that the outline of the later view also was worked out in classical antiquity, actually a little before the time of Aristotle. There is, therefore, a problem about why this picture, which has now ousted its rival, was put aside for so many centuries in favor of the other.

Since there is ambiguity about which was earlier and which was later, I shall now adopt different names for these two world pictures: borrowing from the title of a book by Alexander Koyré,[1] I shall call the one that has now been supplanted 'the Closed World,' and the other 'the Infinite Universe.' Throughout this book, I shall try to preserve the distinction between 'world' and 'universe' in the use of these names: I shall use the word 'universe' to refer to all that there is, and 'world' or 'cosmos' to refer to an organized system of natural parts, usually centered, literally or metaphorically, on the inhabited earth. In the Infinite Universe there may be many worlds, in this sense; but the Closed World, according to its adherents, is identical with the universe.

The aim of this book is to give as exact an account as possible of these two pictures of the natural world and its contents as they were presented by philosophical and scientific authors in classical antiquity. It would not, however, be particularly interesting or enlightening merely to list the features of each and note the points of difference between them. Each was offered as a framework for explanations of the natural world, and each claimed to be better – that is to say, more consistent and more satisfying – than the other. Each was defended by philosophers, who attempted to persuade their readers or pupils by making a reasoned case, not (as sometimes happened in later times) by appealing to human or institutional authority or to divine revelation.

What is important, then, is to examine critically the arguments and counter-arguments put up in favor of each side; and that is what I shall try to do. This attempt is inevitably handicapped by a major difficulty, which it is as well to mention at the start. Classical Greek and Latin texts sur-

[1] All modern works referred to in the text are listed in the bibliography, pp. 201–10.

vived until the invention of printing only if handwritten copies were made and remade throughout the intervening period. There are a few exceptions to this rule, it is true, in the ancient papyrus texts that have luckily been preserved because they lay hidden for ages in some peculiarly dry or airtight place. But it is very rare that these provide more than supplementary evidence, in the form of small fragments. The major part of the inheritance of classical literature was written on papyrus, copied onto parchment, and copied again and again until printed texts were made from these copies. Copying by hand was expensive, and each generation of scholars copied only those texts that they valued highly. Aristotle – '*the* philosopher' to the medieval world – dominated the field of natural philosophy for many centuries; and the result was that the advocates of the alternative cosmological system of the Infinite Universe, particularly Democritus and Epicurus, were despised, neglected, and lost.

There is, therefore, an enormous preponderance of evidence about the system of the Closed World. Some of Plato's dialogues are relevant, especially the *Timaeus*. We have almost all the important works of Aristotle on natural philosophy, and substantial portions of the works of his pupils. On the other side, the major works of Democritus and Epicurus are lost, apart from some tantalizing and difficult fragments. For information about them, we are chiefly dependent on three types of sources: criticisms aimed at Democritus by Aristotle and his pupils; three open letters of Epicurus preserved in the tenth book of Diogenes Laertius' *Lives of the Ancient Philosophers*, and the great epic poem by the Roman follower of Epicurus, Lucretius' *De rerum natura*. Even in the specialized field of biological science, as we shall see later, the same imbalance of evidence occurs. Galen, who was on the side of the Aristotelians, survives in vast quantities, but his physiological and anatomical opponents are known to us chiefly through his criticisms of them.

Before we begin looking at the history of these two world pictures, it will be as well to outline the tenets that particularly characterize them. The names that we have adopted for them, 'the Closed World' and the 'Infinite Universe,' pick out one antithesis among many that separate the two theories.

First, we should notice two features that we may label respectively *evolution* and *permanence*. Within the Infinite Universe, worlds – including our world – grow and decay. There are speculative attempts to describe various stages in this process, from the first formation of the great world-masses (traditionally earth, water, air, and the fiery bodies in the heavens) to the origin of complex living forms, including man, and the subsequent return of the world to its pre-cosmic condition. We can legitimately think of these attempts as theories of *evolution*, although there are vast differences between them and the modern biological theory that has appropriated that name. On the other hand, the Closed World does not evolve. We can distinguish three varieties of the theory in classical antiquity. According to Plato, the world was created by a god. Plato may not have meant this literally, as we shall see when we study his theory in a later chapter, but he was assumed to have meant it literally by many in antiquity. According to Aristotle, the world had no origin and will have no end. It has always been and will always be as it is now. According to the Stoics, the world does have a beginning and an end, but at its end, after a period during which it is consumed by fire, it returns again to its original state and repeats its history in an endless cycle. All three variations have one feature in common, which distinguishes them from the evolutionary theory. The forms – that is to say, the natural kinds – which are realized in the world are *permanent*. According to Plato's creation story, the god who made the world modeled it on eternal forms, which he did not create. Aristotle held that the form of the cosmos as a whole, and the kinds of minerals, vegetables, and animals that now exist are all eternally the same. The Stoics claimed that a 'seed formula' (*spermatikos logos*) for each natural kind was preserved through the periodic conflagrations to start each new world off on precisely the same paths as the old one. In the Infinite Universe, on the other hand, since each world has a beginning in time, there must be an origin of each natural kind found in the world; and since there is no room in the theory for any patterns or models of the natural kinds existing in separation from the world, each natural kind has to be seen as growing or evolving from less complex elements.

From ancient times until the present day, this difference between the two kinds of world picture had the most profound effects. One side is called on to *explain the origin* of complex forms, out of simple elements; the other side is not. It has proved to be the most obstinate problem. As we shall see, Lucretius' attempts to show how the earth could of its own accord produce vegetables, then animals, and even human beings, are almost ridiculously naive. Yet, after centuries of experimentation, the life sciences have not moved far beyond speculation on the origins of life, and their speculations sometimes read astonishingly like passages from Lucretius.

A further major difference can be summed up in the words *mechanism* and *teleology* (I shall say more about these terms in the next chapter). Democritus and the Epicureans sought to explain everything that goes on in the natural world mechanically – that is to say, by reducing it to matter in motion. Indeed, they were mechanists of a particularly austere kind, since they recognized no ultimate change in matter except locomotion, and no interaction between pieces of matter except collision. Every change – even life, and thought – was to be reduced to changes in the motion of particles of matter through space, brought about by collisions with each other (although we must note a partial exception in the Epicurean theory of the 'swerve,' of which more later). Plato, Aristotle, and the Stoics, on the other hand, regarded such explanations as incomplete. A truly satisfying explanation, in their view, should include a statement of the end or goal (*telos* in Greek) to which the explanandum is moving. Although this kind of explanation features predominantly in our talk about human actions, we often extend the same kind of explanation to animals and even plants. So we may say that in this respect the Aristotelians treat the natural world as an organism – Plato indeed says the cosmos is made on the model of a living being (*zöon*) – whereas the Democritean theory uses the model of inanimate matter.

The model of a living organism is often complicated, however – as in Plato's *Timaeus* – by imagery drawn from human arts and crafts. Plato's cosmos is *made by a craftsman*, on the model of a living creature. The directiveness of nature, therefore – that feature of it that makes teleological explanation appear appropriate – could be treated in two ways, both of which are found in the Closed World theories of the classical period. It is sometimes treated as an unexplained feature of natural objects, as a *datum* in the world. The goal for each natural object or process must then be identified, and the rest of the phenomena must be shown to be related to it, but the fact that there *is* a goal is not further explained. This is the type of teleology preferred by Aristotle. Alternatively, the directiveness of nature may itself be explained, as being purposed by a planning mind. Each teleologically explained feature of the world is thus treated as having been chosen for the best, by Providence, or by a beneficent god. This kind of teleology was foreshadowed by Plato and was developed to the full by Stoics and Christians. Both, of course, were rejected by the advocates of the Infinite Universe: the notion of Providence, especially, was a favorite target of Lucretius.

The nature of matter is another subject on which the two cosmologies were opposed to each other. The Democriteans and Epicureans were atomists: they argued that the matter of the universe consists of invisibly small particles, each one being uncuttable (*a-tomos* in Greek), indestructible, indeed totally unaffected by change of any kind except locomotion. Atoms exist and move in a void space that has no center and no boundaries. On the other side, Aristotle and the Stoics (we must postpone consideration of Plato, whose position was not so simple) denied the possibility of a void space. According to the Aristotelian theory, the cosmos is an uninterrupted material continuum, of spherical shape, filling the entire universe.

Perhaps the most important consequence of this last difference, between atoms and the continuum, lies in the theory of motion. It is obvious at once that the two theories must be at odds with each other on the question, because of their opposite views about the void. For the Atomists, the model was of bodies moving freely without any hindrance or pressure except when one collided with another. For the Aristotelians, all motion, except that of the stars at the circumference of the cosmic sphere, was of the nature of swimming through a medium. The medium might be more or less thick, and so motion might be more or less impeded, but the interference of the medium never reached zero. Consequently, *continuing* motion always needed an explanation in this system, whereas the Atomists had to explain only *changes* in motion. At the deepest level, this led to a difference of very great importance: it meant that the Atomists could dispense with gods altogether in their cosmology, whereas the Aristotelian cosmos needed a god to keep it moving. Historically speaking, no feature of the two rival systems had greater significance for their fortunes in the centuries following the classical period.

A difference of equally great historical importance separates the two theories on another problem concerning motion. If we take our observations at face value, there appears at first sight to be a radical difference between the natural motion of objects near the earth's surface – falling stones, or rising flames – and the circular motions of the sun, moon, and stars. Aristotle, believing the cosmos to be eternal, could hardly do other than regard this difference itself as being a part of the unchanging structure, a necessary feature of the nature of things. He posited a radical split between the heavens and the region of the cosmic sphere that lies inside the heavens. The heavens, according to his theory, were made of a special kind of matter, endowed with a natural capacity for motion but otherwise incapable of change. Everything near the center, on the other hand, consisted of matter that tended by its nature to move in straight lines either towards the center or away from it; and this earthly matter was involved in a constant process of interchange, from earth to water to air to fire, and vice versa. In place of this radical dualism, the Atomists posited a single matter – the invariant atoms. They made no basic dichotomy between the circular motions of the heavenly bodies and the rectilinear motions of the earthly elements, but worked out a single theory that aimed to explain both. The model was the vortex. They observed the effects of whirlpools and whirlwinds, natural examples of the conversion of rectilinear motions, such as the normal motion of a wind or the current of a stream, into circular patterns, and used them as models to explain how the rotations of the heavens might have been generated from atoms moving in straight lines in the void. Since they believed that the whole cosmos was itself nothing but a perishable compound, they were not faced, as Aristotle was, with a contrast between the eternal and unchanging heavens and the corruptible beings of the earthly realm. A plant, a man, a moon, or a sun – all had their due span of life, at the end of which they would crumble into their component atoms.

The differences I have sketched are all concerned directly with the natural world and they form the greater part of the subject of this book. We shall not be so directly concerned with two other controversies between these systems, which should nevertheless be briefly mentioned here. Both concern the nature of man, and morality. It was natural enough that the Atomists, since they maintained a thoroughgoing materialism, should hold that there is no life after death: the human soul is a temporary collection of atoms that will be dispersed at death, leaving nothing of the human personality in continued existence. Christian philosophers of course rejected this as anathema and turned above all to Plato for an account of the human soul that they could accept. Aristotle was more ambiguous on the subject, but at least he rejected the materialism of the Atomists and allowed that one function of the soul – an aspect of its rationality – might exist independently of the body.

Christians found as much to be offended by in the hedonism of the Greek Atomists as in their view of the psyche, and greatly preferred the moral doctrines of Plato, Aristotle, and the Stoics. The theory of the Closed World found in nature an eternal organic structure, in which man had a peculiarly honored place at the top of a hierarchy of living forms. The view of the cosmos as an organism suggested that each living form must have a function in the system, just as each of the organs does in a living creature. The function of an organ must be discovered by looking at what it does uniquely, or what it does best. If we look at man's position in the world in this light, his capacity for reasoning appears to be his distinguishing character. Hence, it seemed that the exercise of reason must be man's moral goal; and Plato, Aristotle, and the Stoics all adopted this view of human morality, with some variations of detail. The Atomists, however, saw no such eternal structure in the world. There was nothing given *a priori* by the nature of man's role in a hierarchical order: everything was to be discovered by experience. Pleasure and pain are inescapable experiences and naturally present themselves as the primary motivating agencies; they require no argument, no metaphysical stance, to justify their demands. All that is required (but it is a great deal) is a refinement of man's responses to his own sensations.

We have a number of different areas of thought, then, in which we can distinguish a fundamental opposition between the two world systems; and we shall explore each of these – or at least those concerned directly with the physical world – in the following chapters. But there is also an interesting question about the relation between these different areas. We must ask whether the adoption of a position in one area entails the adoption of a particular position in each, or any, of the others. Is it logically possible, for example, to hold that the universe is a finite, closed system and at the same time to be an atomist? If the relation is less tight than logical entailment, is there nevertheless some weaker connection? We must also shift the question from logic to history and ask whether there were in fact any eclectics in the field of natural philosophy. These last questions will concern us more in the second volume than in the first.

2 The judgement of Socrates

The earliest surviving record of a clash between the two cosmologies is in Plato's *Phaedo* – perhaps an unexpected place to find it, since the *Phaedo* is not about the natural world, but about the human soul, its destination after death, and the implications of immortality for the life of man or earth. And yet Plato's discussion of the issue gains an important dimension from its unexpected context, as we shall see, although he himself does not explicitly draw attention to the point.

Socrates is sitting in an Athenian prison waiting for the death sentence to be carried out. He talks with friends and explains the ground for his confidence that a man's soul survives his death. They listen and are convinced – but two of them express lingering doubts. The first doubt is quickly disposed of, but the second, says Socrates, is more troublesome. The first part of his attempt to allay it contains the critique of other philosophers that interests us. We shall return later to the doubt itself and the role of this critique in putting it to rest.

In his youth, Socrates says, he was an avid student of the philosophy of nature. 'It seemed to me a superlative thing – to know the explanation of everything, why it comes to be, why it perishes, why it *is*' (96a).

The kind of question he considered then, he says, was whether the growth of animals came from a fermentation of the hot and the cold, and whether the blood, or air, or fire, is the organ of thought, or rather the brain, which furnishes the senses, from which come first memory and opinion and then, when these are stabilized, knowledge. But he decided that he had no natural talent for this inquiry, because it left him in confusion and, indeed, shook his confidence in what he formerly thought he knew.

He describes this disaster in strange language; it must contain an element of parody, but presumably it also contains hints of Plato's literal meaning. He used to think the explanation of a man's growth was eating and drinking: the material for the tissues of the body was extracted from the food, so that bone came to be added to the existing bone, and so on. Thus, he thought, a little bulk came to be greater, a small man came to be larger. And when a large man stood beside a smaller man, he was larger, he used to think, *by a head* (he uses the dative case, which can express either the measure of difference, or the cause), and he thought ten was larger than eight because of the additional two, and so on.

But later – having been exposed, presumably, to the philosophy of nature – he was so far from having the explanation of these things that he did not even understand, when one is added to one to make two, whether it is the first one or the second that has come to be two. When they were separate, each was one; now they have joined together, they are two. Is their meeting the explanation of their coming to be two? But sometimes it is separation that produces two. How can opposites, like meeting and separation, be explanations of the same thing?

A ray of light shone on him in this darkness when he heard a reading from a book of Anaxagoras, the fifth-century philosopher of nature: 'It is Mind that puts everything in order, and is the explanation of everything' (97c).

This seemed to Socrates to be on the right track.

> I thought it was somehow right that Mind should be the explanation of all things; if this is so, I thought, then Mind puts everything into order and arranges each thing in the best possible way. So if anyone wants to find the explanation of why a thing comes to be, or perishes, or *is*, what he has to find out about it is how it is *best* for it to be, or act, or be acted on.

So he expected Anaxagoras would go on first to tell him whether the earth is flat or round, and then to recount

> ... the explanation and the necessity, mentioning what is better, and that it was better for the earth to be so.

He expected similar explanations of the earth's position in the middle of the cosmos, and the movement of the sun, moon, stars, and planets; in each case he expected to hear why it is *better* that they move as they do.

> When he said that everything had been put into order by Mind, I did not suppose he was advancing any explanation other than that it is best that things should be so; and so when he gave an explanation for a particular thing or in general for everything, I thought he would give an account of what is best for the particular thing, and of the general good.

This expectation, 'this wonderful hope,' as Socrates calls it, was disappointed. Anaxagoras made no use of Mind as an explanation of the world order, and instead explained things in terms of 'airs, ethers, waters, and such nonsense.'

> It seemed to me [Socrates goes on, sitting on his prison bed] very like a case in which someone says that Socrates does what he does by virtue of mind or inten-

tionally, and then, trying to give explanations for all the things I do, says that I am sitting here because my body is composed of bones and sinews, and the bones are rigid and connected by joints, while the sinews, which can be tightened and relaxed, embrace the bones along with the flesh and skin which hold them together. So when the bones are raised up in their sockets, the sinews, by relaxing and contracting, make me able to bend my limbs now, and that is the explanation of why I am bent into a sitting position here; and he would give other similar explanations for this conversation of ours, too – sounds, air, hearings, and hundreds of other things of the same kind – omitting to mention the true explanation, which is that it seemed *better* to the Athenians to condemn me, and therefore it seemed *better* to me to sit here and righter to stay and submit to the punishment which they command. By the Dog [one of Socrates' favorite oaths], these bones and sinews would long ago have been in Megara or Boeotia, acting on their judgement of what was best, if I had not thought it righter and better to submit to whatever punishment the city imposes rather than run away. To call such things as bones and sinews explanations is too ridiculous. If you want to say that without having bones, sinews, and whatever else of this kind I have I couldn't carry out what seems good to me, you would be right; but to say that I do what I do – and what I do intentionally, by virtue of mind – because of these things, and not through my choice of what is best, would be a very large piece of lazy thinking. (98c–99b)

This speech makes the nature of Socrates' objection to the philosophers of nature a good deal clearer. When asked for the explanation of some feature of the natural world, or indeed for a comprehensive explanation of the whole cosmos, they reply in terms of material components, instead of purposes or 'what is best.' They explain thinking in terms of blood, air, fire, or the brain; they would explain the results of Socrates' own agonizing decision not to take an opportunity to escape from prison in terms of the components of his body. As he goes on to say, they explain the fact that the earth is at rest by saying that there is a vortex round it, or by saying it is underpropped by air; they do not explain what power it is that arranges these things for the best.

We shall have to return later to the kind of explanation that Socrates goes on to present as being more satisfactory than the rejected one. In the present context, let us merely pick out the features Socrates uses to identify the wrong kind of theory, and the weakness that he finds in it. The opposition is not obviously between an 'Infinite Universe' cosmology and a 'Closed World' theory; we shall see later how the features discussed here by Socrates fit into that opposition.

The kind of theory that he used to study, until it led him into confusion, held that the growth of animals is due to a fermentation of the hot and the cold, or that thought is due to the blood, or air, or fire, or the brain. That is to say, it explained complex biological phenomena by referring them to the action of material components. The weakness that he found in this kind of theory was that it said nothing about what is best. Socrates, condemned by the Athenians, decided that it was best to obey the law and not take the opportunity to escape: it is unsatisfactory to offer as an explanation of the action he took as a result of this decision nothing more than a description of his material components. The older philosopher of nature offered *mechanistic* explanations, whereas Socrates wants *teleological* ones. But these names themselves need some comment.

There is nothing misleading about calling Socrates' preferred kind of explanation *teleological*. It explains things by referring to the *telos*: that is, the complete or perfect end to which they are directed. The most obvious kind of event that is normally, in the twentieth century A.D. as in the fourth century B.C., explained in this way is, of course, the purposive action of human beings; it is an action of this kind – his own action in remaining in prison – that Socrates chooses as an example. The action is explained by his purpose, or the end he intends to achieve – namely, to obey the law and so do what is right. But it is not only purposed events that may be explained in this way; to offer a teleological explanation does not necessarily imply that the event is purposed. Certain biological phenomena are normally explained teleologically. We say that the woodpecker has a long, strong beak *for pecking wood*, or that the spider builds webs *to catch flies*, but we do not necessarily imply that the woodpecker or the spider – or even God – had any intentions in the matter. The characteristics of a teleological explanation are, first, that it refers to some result or end achieved (or normally or possibly achieved) by what is to be explained, secondly, that the result or end can be recognized as being in some way advantageous, and, thirdly, that what is to be explained is said to happen *because of or for the sake of* the result or end, rather than producing it by coincidence or accident. The recognition of the advantage must be present if the explanation is to work, but it is the person who accepts the explanation who must recognize it, not necessarily the agent whose action is to be explained.[1]

Although the term 'teleological' is derived from Greek words, the Greeks did not use it. Aristotle, discussing the kinds of explanation, labeled this one 'for the sake of what?' In medieval Latin, it was *causa finalis*, from the Latin *finis* which, like *telos* in Greek, could refer to a purposed or to an unpurposed end.

The word 'mechanistic' is also derived from Greek, without having

[1] More will be said about teleological explanation in volume 2, in connection with Aristotle's theory. For fuller discussion see (for example) Canfield, *Purpose in Nature*; Beckner, 'Teleology', in Edwards, *Encyclopedia of Philosophy*, vol. VIII; Woodfield, *Teleology* (see the Bibliography, pp. 201–10, for details of works cited).

been used by Greeks. In its opposition to 'teleological' it is hallowed by use, and it is impossible to find a substitute without sounding pedantic. But it may be misleading. It suggests an analogy with machines: originally perhaps clockwork was the paradigm case. The performance of a clock is explained by the size, number, and arrangement of its wheels, together with the pendulum or rocker arm and the spring or weight that keeps the motion going. A mechanistic explanation is one that refers to material components and their motions.

The trouble about using this word to apply to ancient Greek theories is that there were no such machines in ancient Greece. Clockwork is regular, predictable, and continuous. Of course, it was made by a clockmaker and needs winding up, but these features are ignored when we think of it simply as a mechanism. The characteristic Greek machine, on the other hand, was a horse-drawn cart, or a winch, which offers no such example of regularity and continuity. More generally, matter in motion, left to itself, on the whole supplied examples of randomness, disorder, and inconvenience. Farmers, gardeners, the owners of boats, and all householders have a painfully clear picture of the contrast between human purpose and the undirected products of matter in motion. In general, anything that exhibits order is likely to be attributed to human action. This becomes important when it is suggested that the whole world order is a product of matter in motion. The implausibility of the suggestion, in ancient Greece, may be concealed if we use the expression 'mechanistic explanation' unreflectingly. When Aristotle contrasts something that is to be explained teleologically with something to be explained mechanistically he calls the first 'for the sake of something' and the second 'from necessity.' In another context, he contrasts what is purposed with what is 'automatic.' (We shall explain Aristotle's use of these terms later on.) These might give us some justification for speaking of 'necessity' or 'automatism', instead of 'mechanism.' But the first is too ambiguous and the second too esoteric; so it seems best to stay with 'mechanism,' in spite of its faults.

It is striking that in this speech of his in the *Phaedo* Socrates expresses a preference for teleological explanations over the whole range of natural phenomena. He wants teleological explanations not only for purposed actions like his own sitting in prison and for biological phenomena like growth, but also for the shape of the earth and the movements of the sun, moon, planets, and stars. He shows no consciousness of a difference in kind. It has often been said that teleology should play no part in natural science, and that the advance of science, historically speaking, was retarded for many centuries by the demand for final causes. At the beginning of the 'scientific revolution,' Francis Bacon, in his *Novum Organum* of 1620, mounted a full attack on the use of final causes in science – not on the ground that final causes are altogether non-existent and imaginary, but that they are inappropriate to the philosophy of *nature*. If it is true that the demand for teleological explanation was a prolonged hindrance on the progress of science, then we must accuse Socrates of being the first in a long line of obscurantists. But we shall see later that the truth is not so simple and clear-cut.

Teleological and mechanistic explanations are not necessarily incompatible with each other. Socrates makes it clear that he has understood this point when he allows that matter in motion may be allowed to be the *sine qua non* of what is to be explained. Bones and sinews and their movements will not suffice to explain his sitting in prison, but *without* bones and sinews and their motions, he would not be able to sit in prison. Matter in motion is necessary, but relatively uninteresting. Aristotle's biological science is built on the foundation of this relationship between the two kinds of explanation.

We may return now to the context in which Plato introduces Socrates' autobiographical account of his early studies in natural philosophy. There is no need to enter into the controversy about the historicity of this account.[2] The evidence on that is conflicting. According to Aristophanes' satirical portrait of Socrates in his comedy *The Clouds*, Socrates taught natural philosophy and conducted his own researches (of a kind – for example, how many times its own length does a flea jump?). According to Xenophon's *Reminiscences of Socrates* (I.4 and IV.3), he taught his pupils a teleological theory of the natural world. But in Plato's *Apology* – Plato's account of the speech Socrates made in his own defense at his trial – he denies emphatically that he has ever been heard discoursing about nature. Without asking, then, whether the real Socrates was ever a philosopher of nature, we can still raise the question of what Plato is up to in this part of the *Phaedo*.

The problem presented by Cebes to Socrates in the dialogue is this: what reason have we for denying that the human psyche is durable enough to outlive many bodies, but nevertheless perishes ultimately, like a coat that can be handed on from one wearer to another but will be worn out in the end? Socrates meets this objection in a passage over 5000 words long, of which the 'autobiography' is the beginning. By the time he returns to face the objection directly, the critique of earlier philosophers of nature has receded into the background; Socrates does not explicitly point out its relevance to the problem of the soul. Moreover, his account of the psyche

[2] See, for example, Guthrie, *HGP*, vol. II, pp. 421–5.

is not overtly teleological; and since it is teleology that he opposes to mechanism in the passage we have been studying, this may suggest that the whole critique of mechanistic theories is not directly relevant to the problems of the psyche and supplies only a background to the problem. The morals of Plato's dialogues, however, are not all spelled out in so many words. Perhaps we can legitimately look for an implicit point here. If so, the point must be just that human life cannot be explained adequately by talking about matter in motion. The psyche is not the same kind of thing as the 'airs, ethers, waters, and other such nonsense' of the natural philosopher's theories. To give an account of it requires a different kind of thinking altogether.

We can reasonably interpret his criticism of early natural philosophy not only as a demand for teleological explanations instead of mechanistic ones in natural philosophy itself, but also as a hint that the whole enterprise of the natural philosopher was aimed in the wrong direction. In fact, Socrates and Plato thought much more about man than about nature: the central problem became the problem of human life – how can we know what is the right way to live? When Plato returns to the study of nature, in the *Timaeus*, the cosmology that he presents is focused on man much more closely than the Presocratic theories. As usual in the Closed World, man is at the center.

But we must now go back to earlier history, and investigate the origin and growth of the mechanistic theories criticized by Socrates.

3 The beginning in Miletus

There are many good histories of early Greek philosophy, and it is not necessary to the purpose of this book to attempt to rival them.[1] We must try to pick out the themes outlined in the earlier chapters, without encumbering the narrative with too much detail. The reader must be warned, however, that this selectivity is certain to be controversial. The evidence for the theories of the sixth and fifth centuries B.C. is scrappy and ambiguous: we lack the context of the short quotations that survive – the longest consecutive fragment is sixty-six lines of verse. It is just possible that we have the whole intent and direction of some fragments wrong. That is not likely, because the tradition has been subjected to the most careful criticism by many generations of scholars and philosophers; but there is no general consensus on some important questions, and the possibility of misinterpretation is open. This is as a rule not true in the case of philosophers like Plato and Aristotle, whose work survives either completely or in bulk.

The main theme to be picked out is the one emphasized by Plato, in the passage of the *Phaedo* that we discussed in chapter 2: the explanation of the world in terms of matter in motion. We must begin, however, briefly, at the beginning: that means, with the three philosophers who lived in the sixth century B.C. in the prosperous Greek colony at Miletus, on the coast of Asia Minor – Thales, Anaximander, and Anaximenes.

3.1 *The origin of the world: matter alive*

Before philosophy, there was myth; but to write accurately about this relationship requires a delicate hand. The categories of 'myth' and 'philos-

[1] The Greek texts of the fragments and testimonia of the Presocratic philosophers are collected in Diels–Kranz, *Fragmente der Vorsokratiker*, cited here as DK.

There is a full account of the Presocratics in Guthrie, *A History of Greek Philosophy*, vols. I–II, cited here as *HGP*; less full in Robinson, *An Introduction to Early Greek Philosophy*; good chapters on 'The Forerunners of Philosophical Cosmogony' and on the Milesians in Kirk, Raven and Schofield, *The Presocratic Philosophers*. Barnes' *The Presocratic Philosophers*, is lively and stimulating, but rather idiosyncratic.

For an extreme statement of the sceptical view of the Milesians, see Dicks in *JHS* 86 (1966), and his *Early Greek Astronomy to Aristotle*, ch. 3. There is a very clear and full account of the evidence for Milesian theories in Kahn, *Anaximander and the Origins of Greek Cosmology*; he also responds to Dicks in his article in *JHS* 90 (1970).

ophy' are modern ones. We do not know how Thales described his own intellectual activity, but it is unlikely that he called it 'philosophizing.' We must not exaggerate the difference between Thales and his predecessors, as if suddenly, conscious of what he was doing, he built and launched a new kind of boat and sent it off to pick up Plato, Aristotle, Kant, and Wittgenstein. At the same time, it is surely correct to see something new in the very fact – well attested as it is – that Thales, Anaximander, and Anaximenes each produced a theory about the nature of the world, and each in turn differed from his predecessor. Although they may not have constituted a 'succession' or a 'school' of philosophy, it seems undeniable that Anaximander proposed an alternative theory such that his own and Thales' ideas could be assessed comparatively, and reasons given for preferring one to the other; and that Anaximenes treated Anaximander similarly. They began, then, a tradition of critical discussion; and that is something not generated by myths.[2]

One idea that they shared with the myths is that it is appropriate to talk about the *origin* of the world and its parts. There is an important difference, however, in the nature of the agents who bring about this origin, and in the mode of their operation. The myths tell stories of human or quasi-human agents, moved by human emotions to perform actions that are modeled, quite evidently, on human behavior. The Milesian philosophers began the move away from human categories, towards matter in motion: the kind of motion that they attributed to matter was far from mechanical, as we shall see, but their work must count as the first step towards the mechanization of the world picture.[3]

Two quotations from myth will help us to make this contrast:

> Valiant Marduk
> Strengthened his hold on the vanquished gods,
> And turned back to Tiamat whom he had bound.
> The lord trod on the legs of Tiamat,
> With his unsparing mace he crushed her skull.
>
> He split her like a shellfish into two parts:
> Half of her he set up and ceiled it as the sky,
> Pulled down the bar and posted guards.

[2] That the Milesians originated the tradition of critical discussion is maintained by Sir Karl Popper, *Proceedings of the Aristotelian Society* 59 (1958–9).

[3] For the relation between Near Eastern myth and Greek philosophy, see especially Frankfort, *et al. Before Philosophy*; Cornford, *Principium Sapientiae*; Hölscher, *Hermes* 81 (1953); and West, *Early Greek Philosophy and the Orient* (rather hasty and erratic). For the text of Near Eastern cosmogonical myths (in English translation), see Pritchard, *Ancient Near Eastern Texts Relating to the Old Testament*.

He bade them not to allow her waters to escape.
He crossed the heavens and surveyed the regions.

These lines are from the Mesopotamian creation epic, known as *Enuma Elis* (Pritchard, *Ancient Near Eastern Texts*, Tablet IV). Tiamat is a goddess of the water of the sea, who is killed and split in half by Marduk, to make an upper half and lower half. The upper half becomes the sky, which is then decorated by Marduk with likenesses of the gods – the constellations.

The Greek myth of the origin of the world is a story of a birth, rather than a creation. It survives in Hesiod's *Birth of the Gods* (*Theogony*):

Hail, offspring of Zeus, grant me lovely song:
Make famous the holy family of the immortals who live for ever,
Who were born of Ge [Earth] and starry Uranus [Heaven],
And dark Night, and those whom salty Sea brought up.

. . .

First of all was Chaos born; and then
Gaia [Earth], wide-bosomed, a sure, eternal home of all
The immortals who possess snowy-peaked Olympus;
Also misty Tartarus, in a nook of the wide-wayed earth;
And Love, most beautiful among the immortal gods,
Looser of limbs, for all gods and all mankind
Conqueror of the mind in their breast and of wise counsel.
From Chaos, Erebus and black Night were born;
From Night again there were born Aether and Day
Whom she bore, conceived through mingling in love with Erebus.
Gaia [Earth] bore first the equal of herself,
Starry Uranus [Heaven], to cover her entire,
That there might be a sure, eternal home for the blessed gods.
(*Theogony* 104–28)

The essence of the Milesian theories, which we must now compare with the dramas of the myths, can be given in a single word: hylozoism, the doctrine that matter as such has the property of life and growth. They all offered a view of the world as having grown to its present form from a kind of seed. The seed was a homogeneous material substance. According to Thales it was moist, like the semen of animals; according to Anaximenes it took the form of the air that is breathed by living creatures. These two appear to have held that the seed-substance was itself primitive and underived: once upon a time there was nothing but moisture – or air – in an undifferentiated state; spontaneously, having the capacity for life in it, it grew by stages into an ordered structure of earth, sea, and sky, which in turn gave birth to vegetable and animal life. The middle member of the trio, Anaximander, differed from Thales and Anaximenes in proposing that the seed was not itself primitive, but was secreted from the primitive material body, which he called simply 'the Boundless.'

The evidence for the statements of the last paragraph is not to be found in direct quotations from the Milesian philosophers but in later Greek summaries of their doctrines. One or two examples may be useful to give an idea of the nature of this material. First, a passage from Aristotle:

> Of the first philosophers, most thought the principles that were of the nature of matter were the only principles of all things. The element (*stoicheion*) and principle (*archê*) of the things that exist, they say, is that from which all things have their being, from which they come into being in the first place, and into which they perish at the end, the substance of this persisting while it changes its properties... But as to the number and form of such a principle they do not all say the same. Thales, the originator of this kind of philosophy, says it is water (hence he declared the earth floats on water), taking this idea, one may suppose, from the observation that the nutriment of all things is moist, and warmth itself [*sc.* the warmth of living things] comes from moisture and lives by means of moisture – taking the idea, then, from this, and from the fact that the seed of all things has a moist nature; and water is the principle of the nature that moist things have... Anaximenes and Diogenes [*sc.* of Apollonia, a much later philosopher], on the other hand, make air prior to water and the principle of the simple bodies, whereas Hippasus of Metapontum and Heraclitus of Ephesus choose fire, and Empedocles the four, adding earth to the other three [*sc.* water, air, and fire]. (Aristotle, *Metaphysics* 1.3, 983b6–984a10, excerpted)

The next passage is from a work called *Stromateis* or *Patchwork Quilt*, quoted by Eusebius and alleged by him to be by Plutarch:

> They say that Thales, first of all, posited water as the *archê* of the whole universe, since from it all things have their being, and into it all things pass.
>
> After him Anaximander, who was a companion of Thales, says that the Boundless is the whole cause of the generation and destruction of the whole. Out of it, he says, the heavens were separated off and in general all of the worlds, being boundless. He declared that destruction takes place, and, much earlier, generation, from boundless ages, when they all come round in cycles(?). He says that in shape the earth is cylindrical, while its depth is one third of its breadth. He says that a certain seed of hot and cold, out of the eternal, was secreted at the generation of this cosmos, and a certain ball of flame, out of this, grew around the air that surrounds the earth, like bark around the tree. When this ball was broken apart, and closed off into rings, the sun and moon and stars came into existence. Further, he says that in the beginning Man was generated from living things of different form, from the fact that other things soon get food for themselves, whereas Man alone requires lengthy nursing; hence he would not have survived, if he were of this description right from the beginning.
>
> They say that Anaximenes said the *archê* of the universe is air, and that this is

boundless in magnitude but bounded in the qualities associated with it. And all things are generated according to a certain condensation and then a rarefaction of this; whereas motion is eternal. He says that when the air hardens like felt, the earth is first created, very flat – hence, reasonably enough, it rides on the air. And he says that the sun and the moon and the other heavenly bodies have the beginning of their generation from earth. At any rate, he says the sun is earth, and from its swift motion it gets very adequate heat. (Eusebius, *Praeparatio Evangelica* 1.7 = *Stromateis* 1–3)

The dominant idea that emerges from these summaries is that of the progressive growth of the cosmos, starting from an undifferentiated mass. First there is something called the *archê*, which means 'beginning' or 'starting-point' or 'principle'; and although the Milesians differed about its identity, they agreed that it was in the first place unified and homogeneous. Its function in their theory was to produce out of itself a plurality of different things, which would subsequently become the cosmos, and apparently to receive back into itself all the materials of which the cosmos was composed, after its dissolution.

Such a theory presupposes the conservation of matter, in some sense. 'Nothing comes into being out of nothing, and nothing passes away into nothing.' This became a slogan of Greek natural philosophers, of all persuasions and all ages; even those philosophers who thought the world was created by a Creator God described him as shaping the cosmos out of pre-existing materials, not as conjuring it out of nothing. In addition to asserting the conservation of matter, the text we have quoted from pseudo-Plutarch also hints (it can hardly be said to do more) at a theory of cycles, such that the world is said to emerge from the *archê*, return to it again, and re-emerge once more, in a recurring pattern. Since the evidence is scrappy we cannot attribute this idea to the Milesians with confidence; but later philosophers in the Greek tradition adopted it unambiguously. I shall have more to say about it later.

But the Milesian notion of the conservation of matter needs some further clarification. The principle in their theory, says Aristotle in the *Metaphysics* passage just quoted, is 'that from which they come into being in the first place, and into which they perish at the end, the substance of this persisting while it changes its properties.' Some have taken this to imply that in Thales' view, for instance, the water that constitutes the *archê* of the world continues to exist as water throughout the life of the world; so it would be reasonable to say of everything in the world, at any time, 'it is really (basically, fundamentally) water.' Probably this gives a misleading gloss on the theory, with its suggestion of an underlying, hidden reality, contrasted with the perceptible appearance. Aristotle is using his own terms, 'substance' and 'properties', to describe an older notion. The persistence of the *archê* is probably no more than its spatial identity through time; just as we can point to a young oak tree and say 'that is the acorn that I planted,' without wanting to say 'it is still an acorn,' so we can think of Thales' developed cosmos as being a new state of the one original *archê* without implying that it is still water.

The stages of the emergence of the world from the *archê* cannot be distinguished with much clarity. Eusebius, writing about Anaximander, speaks of a 'seed of hot and cold,' or more literally, 'a thing productive of hot and cold.' The distinction between substances and their properties lies in the future: we should think of 'hot' and 'cold' as naming masses of matter characterized by just these properties. The hot and the cold should be thought of as either identical with, or about to produce, the heavens and the upper air, and the earth and sea, respectively. The first offspring of the *archê* – or at most, the second – are the great world masses. But are we to say they are the first, or the second? Some have argued that this reference to the 'opposites,' hot and cold, is too abstract and schematic for this early stage of cosmological thought. Others have pointed out that the hot and the cold, under those very names, are well attested in early medical writings and in Presocratic philosophy of not much later date – not, of course, as adjectival qualitative properties, but rather as 'powers' (*dynameis*), beings capable of coming and going, and of imparting their own nature by their presence or of making way for their opposite by their absence. In any case, it is clear that *opposites*, under some description, played an important part at least in Anaximander's theory, as we shall see.

The main explanatory model chosen by the Milesians for their cosmogony was a biological one: growth from seed. Contemporary understanding of biological processes had certain deficiencies that made this choice of model more plausible. First, it was widely thought that there are instances in nature of spontaneous generation, without any parentage: certain fish, testaceans, and insects are quoted as examples.[4] Moreover, the role of the female in sexual reproduction was not understood: it was thought that she provided merely a suitable environment, a seed-bed, for the development of seed, or at most an inert material component to be worked on by the male seed. The active power of initiating growth lay in the semen alone. Some sort of explanation of this power was sometimes found in its moist nature, since the relation of moisture to life was abundantly obvious to anyone with experience of growing crops in the Aegean or Near Eastern regions: hence, the primitive moisture of Thales' theory.

[4] Texts are listed in Balme, *Aristotle's De Generatione Animalium I*, p. 128.

Alternatively, and less obviously, the secret could be located in its foamy character, interpreted as showing an airy or breathy component. Breath is as obviously necessary for life as water. It may be guessed that Anaximenes chose air as the primitive substance both because he supposed it to be Boundless, as the origin of all things should be, and at the same time because it had in itself the same life-giving capacity that Thales' water had.

In the reports of the theories of Anaximander and Anaximenes we find clear hints of a tendency to pick out certain forms of matter as more basic or more primitive than others, and these are seen as being grouped in pairs of opposites. As we have already noted, the primitive seed in Anaximander's theory first produced 'the hot and the cold,' according to one report. Anaximenes is said to have observed that breathing out with the lips relaxed produces breath that feels hot, whereas breath forced through compressed lips feels cold.[5] Transferring this idea to cosmology, he claimed that the rarefaction of the primitive air produced fire, and its condensation produced the colder kinds of matter, wind, cloud, water, and earth.

Even in the sixth century, then, we can find traces of ideas that were later to become canonical. It would be anachronistic to speak of four elements: they were postulated clearly for the first time by Empedocles. But the Milesians had an intuition of four states of matter that we can loosely identify as fire, air or vapour, water or liquid, and earth or solid matter. And they associated these with certain predominant perceptible qualities: hot and cold, dry and wet. The pattern is somewhat obscure and confused in the Milesians: Aristotle later gave it a formal precision by giving a pair of qualities to each kind of matter – fire is hot and dry, air hot and wet, water cold and wet, earth cold and dry.[6]

Although we may allow that the Milesians anticipated fifth-century mechanistic theories in picking out certain forms of matter as being simpler and more primitive than those which later emerged from them, I believe it is probably anachronistic to attribute to them the mechanistic theory of the vortex. Democritus and others certainly postulated the vortex as the explanation of the sorting of particles of matter by shape and size: we shall investigate the idea below (see chapter 10). But the model consorts badly with the biological model of growth. It is true that the vortex was used later to illustrate the growth of an embryo, by the author of the Hippocratic treatise *The Nature of the Child*, which will be discussed in more detail in chapter 10. But that was a technical theory of bio-

[5] DK 13B1 (Plutarch).
[6] For Aristotle's theory, see his *De generatione et corruptione* II.1–8, especially 3.

logical growth, whereas the Milesians appealed to ordinary experience.[7]

3.2 *The dynamics and shape of the world*

For the maintenance of the cosmic order, as opposed to its first appearance, Anaximander appealed to the analogy of human societies. The cosmic order consists of a just balance between opposite interests: the heat of summer is recompensed by the cold of winter, the light of day by the dark of night, the death of one thing by the birth of another.

> Anaximander says that ... whatever things are the case come-to-be out of certain things and pass away into those same things 'according to necessity, for they pay penalty and recompense to each other for their injustice according to the assessment of time.'[8]

Motion and change in the natural world are thus hardly explained, but at least brought into a familiar perspective by being described in language commonly used of the phenomena of living forms, including the behavior of man. At this early stage, there is little point in asking whether the theory is a teleological or wholly mechanistic one. The Boundless, says Aristotle, 'seems to be the beginning of all things, and to encompass all and to *steer* all ... And this [*sc.* the Boundless] is the divine, for it is immortal and indestructible – so says Anaximander, and most of the natural philosophers' (*Physics* III.4, 203b9ff.). If Anaximander thought of the Boundless as an immortal god and as steering the course of the developed world, then one might at once claim that he should be classed among the Providentialists. On the other hand, the biological imagery suggests natural growth, rather than purposive action. Perhaps the metaphor of steering means no more than that the Boundless, which is the origin of the world, starts it on a particular course: its first product is 'the seed of the hot and the cold,' and no doubt the fact that it is just these two opposites

[7] Among those who support the notion of a vortex as part of the theory of Anaximander are Heidel, *Classical Philology* 1 (1906), and Robinson, 'Anaximander and the Problem of the Earth's Immobility', in Anton and Kustas, *Essays in Ancient Greek Philosophy*. Although I agree with them, against most of the commentators, about Anaximander's theory of the earth's stability (see below), I think they are wrong about this. Some counter-arguments can be found in Ferguson, *Phronesis* 16 (1971). But counter-arguments are hardly needed, since (*pace* Robinson, 'Anaximander', p. 117) there is virtually no evidence for a vortex in Anaximander: only a very general reference in Aristotle's *De caelo* II.13, to 'all who hold that the heaven came into being' (295a10), and the word 'whirl around' (περιδινεῖσθαι) applied to the star wheels by Aetius (DK 13A12) without mentioning Anaximander by name.

[8] DK 12B1. This is the only sentence surviving in Anaximander's own words, and it has attracted an enormous amount of literature. There is a handy account of it in Classen, 'Anaximandros', in Pauly–Wissowa, *Realencyclopaedie der classischen Altertumswissenschaft*, suppl. 12, cols. 56–60. See also Kahn, *Anaximander*.

that are the first to grow affects and even determines the future course of the world. Is the *balance* of opposites in the world, achieved when 'they pay penalty and recompense to each other for their injustice,' to be thought of as a mechanism, or as a goal aimed at by nature? There seems to be no way of deciding the issue.[9]

How was our world pictured when it had grown from its original source? The earth is flat. Thales, according to Aristotle, described the earth as being supported by water, like a piece of floating wood – and Aristotle criticizes Thales for not asking how the supporting water is itself held up. It is important to notice the directionality suggested by this picture and the criticism. Water prevents the earth from falling *downwards*, and Aristotle wants to be told why the water does not fall downwards: the suggestion is that the direction of downward fall, as seen above the surface of the earth, continues indefinitely below the earth's surface. A similar picture is attributed to Anaximenes, substituting air for water: 'the earth's flatness is the cause of its remaining in place, for it does not cut the air underneath but covers it like a lid.' The same directionality can be seen in Hesiod's *Theogony*: he tells us that an anvil of bronze – that is to say, something *heavy* – might fall from heaven for nine nights and days, to arrive on the tenth day on earth; and the anvil might then fall from earth for nine nights and days, to arrive in Tartarus on the tenth day. It is a *fall*, all the way. Thales and Anaximenes no longer thought of the earth as having roots in Tartarus, as Hesiod did, but they still sought for a reason why it did not fall.[10]

The theory attributed to Anaximander, however, is startlingly – perhaps incredibly – different. Aristotle says:

> There are some, like Anaximander among the ancients, who say the earth remains in place because of its uniformity [*lit.* likeness, similarity]. For whatever is settled at the center, and is uniformly related to the extremes, should be not more inclined to move either upwards or downwards or sideways. But it is impossible for it to move simultaneously in opposite directions; hence it necessarily remains in place. (*De caelo* II.13, 295b10)

To understand how extraordinary this is, we must introduce a contrast that becomes explicit and important later in this history. According to the Aristotelian theory of the Closed World, the universe has a center, which is the point of reference from which all motion is determined: we will call

9 The ideas involved in Anaximander's treatment of Justice in the cosmos and the divinity of the Boundless are well discussed in Vlastos, 'Equality and Justice in Early Greek Cosmologies' and 'Theology and Philosophy in Early Greek Thought,' both repr. in Furley and Allen, *Studies in Presocratic Philosophy*, vol. I, pp. 56–91, 92–129.

10 Thales, in Aristotle, *De caelo* II.13, 294a28 = DK 11A14. Anaximenes, in Aristotle, *ibid.* 294b13 = DK 13A20. Hesiod, *Theogony* 720–5.

it 'the centrifocal theory.' Heavy bodies naturally fall towards the center, light bodies rise from the center, and the heavenly bodies move around the center. This theory implies that the cosmos is spherical, and it needs only a brief argument in Aristotle's *De caelo* to show that the earth must be spherical too. On the other hand, the Epicurean Atomists (I omit Democritus for the present as being too controversial) pointed out that if the universe is infinite, it has no center. Natural motion, therefore, cannot focus on a center; it must be in *parallel* lines. Since the direction of the fall of heavy bodies on earth at all points is observed to be perpendicular to the earth's surface, this appears to entail that the earth is flat. And the question is raised by this picture, as it is not in Aristotle's system, why the earth itself does not fall downwards through space.

Now in the Milesian story we seem at first sight to have elements of both these pictures. Thales and Anaximenes claim that the earth does not fall because it is supported from underneath, and that claim makes sense only in a framework of parallel motion. But Anaximander is said to have claimed that the earth is in equilibrium because it is at the center, and that appears to make sense only on the centrifocal theory of motion. However, this goes along with a flat earth: he is said to have compared the earth to the round drum of a pillar, of height equal to a third of its diameter. Anaximander is often complimented for his theory of the earth's position: it is called 'a brilliant leap into the realms of the mathematical and the *a priori*'; 'nothing shows more clearly the independent quality of Anaximander's mind.'[11] But I find it very hard to make coherent sense of it. We must press two questions: what exactly could the earth be at the center of, and why should being at that center entail being in a state of equilibrium?

Anaximander is credited by the ancient sources with the theory that the sun and moon and stars are the visible portions of rings of fire, which circle around the earth.[12] At its first growth, the cosmos was enclosed within 'a ball of flame'; this ball was broken up into rings, each enclosed in a covering of air or mist. The covers were punctured in certain places, allowing the enclosed fire to be seen from the earth. The apparent waxing and waning of the moon is due to the blocking and unblocking of the hole in the moon's ring: and eclipses are similarly explained. So if Aristotle speaks of Anaximander's earth 'uniformly related to the extremes,' it can hardly be anything other than these rings that he has in mind.

[11] The compliments to Anaximander are from Kirk, Raven, and Schofield, *The Presocratic Philosophers*, 2nd ed., p. 134, and Guthrie, *HGP*, vol. 1, pp. 99–100. Others who have found the attribution of this theory to Anaximander incredible are *Classical Philology* 1 (1906) and Robinson, 'Anaximander'. I have argued the case at more length, somewhat differently from them, in a paper to be published in my *Cosmic Problems*.

[12] DK 12A10–11.

But why should the existence of these rings have any influence on the possible downward motion of the earth? Aristotle's description makes sense on his own theory, according to which the forces causing upward or downward movement are symmetrical about the central point of the cosmic sphere. But why, in Anaximander's system, should a position on the axis of the star rings count as a reason for the stability of the earth? The evidence gives no hint of a theory of attraction, which might account for it. Nor does it seem reasonable to think of earth as tending to *fall* equally in all directions, since it clearly falls in just one direction whenever we can observe it (supposing the earth's surface to be flat).

I am inclined to think that there is some confusion in the reports of Anaximander's theory. Aristotle's report, on which the whole story depends, is actually about Plato's *Phaedo*: Anaximander is mentioned only in passing, and it is quite unclear how much of the *Phaedo* theory Aristotle intends to attribute to Anaximander. If Anaximander spoke of the uniformity ('likeness') of the earth, perhaps he meant only that it had no inclination to *tilt* one way rather than another and therefore remained afloat. As we have seen, the Milesians were impressed by the ability of objects to remain afloat so long as they were flat. Simplicius, who had access to more direct evidence on the Presocratics than we have, comments on this passage of Aristotle: 'Anaximander thought that the earth remains in place *because of the air that supports it* and because of its equilibrium and uniformity' (*De caelo* 532.13). If Anaximander, like Anaximenes, supposed the earth to be afloat on air, it would make perfectly good sense for him to invoke the 'equilibrium' or 'likeness' of the drum-shaped earth as an explanation of why it does not tilt like an unevenly loaded raft, and so cut through the air, instead of resting on it.

But before leaving this subject, we must return to the 'ball of flame' mentioned by our source as the origin of the star rings:

> He [Anaximander] says that a certain seed of hot and cold, out of the eternal, was secreted at the generation of this cosmos, and a certain ball of flame, out of this, grew around the air that surrounds the earth, like bark around the tree. When this ball was broken apart, and closed off into rings, the sun and moon and stars came into existence. (*Stromateis* 2)

This fascinating passage has contrary implications about the shape of the cosmos. The image of a tree-trunk suggests something roughly *cylindrical*: yet the origin of the star rings is said to be a *sphere*. Which was the view of Anaximander?

The history of Greek astronomy has a clearly distinguishable starting-point, and a well-known end point; the problem is to discern the stages by which progress was made from one to the other. The starting-point is the picture of the world given by the Homeric poems and Hesiod – the picture of a flat earth, with a solid, dome-shaped sky arched over the top of it. The sun, moon, and stars make a daily journey from east to west, in an arched orbit over the earth, and they return to their starting-points in the east, after 'bathing in Ocean,' by circling in some unexplained way *around* the earth. Under the earth is Tartarus: we have already seen a hint of symmetry in the relation between what is under and what is over the earth, in Hesiod's bronze anvil, which takes the same time to fall from sky to earth, and from earth to Tartarus. But there is not yet any symmetry between the visible and invisible portions of the orbits of the heavenly bodies. Some kind of exception to this is provided by the stars, which, as seen from Greek latitudes, never 'bathe in Ocean': the circumpolar stars, whose courses, observed during a whole year, describe complete circles around a fixed point in the sky. But this observation was not immediately generalized to cover the stars whose orbits are not fully visible.

At the later end of the historical development, we have a fully spherical picture. The fixed stars are thought of as moving in concert on the inner surface of a sphere that rotates once a day around its axis, at the center of which is the earth. The paths of the heavenly bodies that move at variance with the fixed stars – the sun, moon, and planets – are analyzed by assuming the existence of invisible spheres inside the sphere of the fixed stars, each rotating on poles, which are themselves carried around by the motion of the next outer sphere in which they are set. The stars do not go 'down,' into Ocean or anywhere else, when they set: 'down' now means 'towards the center' of the sphere.

At what point in history was the first move made towards the spherical model? The notion of star rings in Anaximander appears to provide the answer. An original envelope of flame enclosing the cosmos was closed up in rings of mist, in such a way that the flames were visible through punctures in the mist rings at certain points. The flames visible through these holes are our sun, moon, and stars. Our sources explicitly mention the sun and moon as being explained by this idea: they are among the heavenly bodies that set, and so we may assume that all the stars were believed to complete circular orbits.

It should be noticed that the idea of circular orbits for all the heavenly bodies does not itself entail that the whole cosmos is spherical. Since the stars do not obviously vary in brightness at different points in their orbit, it is a natural assumption that the earth is equidistant from all points in the orbit of each star – that is to say, the star wheels are all rotating on the same axis. But this is consistent with giving the whole cosmos the shape of an egg, or any spheroid. The only objections to a cylindrical shape are that we do not see the stars getting smaller and dimmer nearer the pole, and we do not see a hole, empty of stars, at the top.

In any case, however, it seems unlikely that Anaximander attached importance to the overall shape of the cosmos as a whole. We are told that he made the circle of the sun larger than that of the moon, and the moon's larger than the stars'. No dimension is given for the stars' rings, but the moon's is said to be 18 times the size of the earth, and the sun's 27 times; it is a reasonable guess that 9 was the figure for the stars. The motivation for this pattern is evidently quite different from the geometrical inspiration of the spherical cosmos: it seems likely, indeed, to have arisen from a knowledge of Iranian ideas, according to which the intervals measured by earth, stars, moon, sun represent steps in the ascent of the soul towards the light.[13] The mythical background is confirmed by the close parallel with Hesiod's nine-day interval between heaven and earth, and earth and Tartarus, mentioned above.

If the notion of star rings was an innovation made by Anaximander, it would provide a reason for his abandoning Thales' notion that the earth floats on water. The path of the stars took them under the earth: presumably their way would lie through air or aether under the earth as above it. In that case, it would have to be explained why the earth does not fall through the air underneath it, as we see that any piece of earth falls if there is nothing but air under it. My suggestion is that Anaximander believed it to ride on the air because it is flat and evenly balanced.

3.3 *Anaximander and infinity*

Before leaving the Milesians, we must say something about their notion of infinity. Anaximander spoke of 'the Boundless' (this translation of *Apeiron* is adopted here to avoid the excessive precision of 'the Infinite') from which there emerged the first differentiated elements at the origin of our world. Superficially, this suggests that he is to be classed among the advocates of the Infinite Universe as opposed to the Closed World. But as with the dichotomy between mechanism and teleology, it would probably be anachronistic to put him firmly into either camp. The evidence, however, is confused. Perhaps it will be best to mention my own view first, and then discuss briefly some considerations that seem to tell against it.

I think we do best to regard the postulation of the Boundless simply as a

[13] See Burkert, *Rheinisches Museum* 106 (1963).

decision not to put questions beyond a certain point. A originates from B, B from C, C from D . . . Anaximander puts a stop to this series by asserting that there is something that *has* no origin, but *is* the origin: he simply dismisses the question 'But what is *its* origin?' as illegitimate. We can give that question different senses, and arrive at a different notion of the Boundless each time. If we give it a temporal sense, the Boundless takes on the aspect of eternity. If we think of spatial extension (as when we say that Italy *begins* at the border of Switzerland) then the Boundless is spatially infinite. To call something by the name of any of the opposites (e.g. hot or cold) is to mark it off as being distinct from other things; to say something is Boundless is to say it is not marked off from anything else, by heat or cold or difference of density or any other character.

I take it, then, that Anaximander thought of our world as embedded in a surrounding mass of an indeterminate substance that had always existed and stretched without limit or variation in all directions.[14]

There is, however, an ancient tradition, probably going back to Theophrastus, that Anaximander postulated 'innumerable worlds.' Now, as we shall see, it is an integral part of the theory of the Infinite Universe that a world is simply a mortal compound like any other, and that there is no limit to the number of worlds that may be in existence, now or ever. If this theory is to be ascribed to Anaximander, then we shall have to accept him as the founder of the Infinite Universe theory. But it is doubtful whether the reports ought to be taken at face value. This is not the place for a critical examination of the evidence. It must suffice to say that there are reasons for thinking that Anaximander, who postulated the *Apeiron*, may have been simply confused with the Atomists, who postulated the *Apeiron and* innumerable worlds. If Anaximander wrote something about 'infinite *ouranoi*,' he perhaps meant that there was not just one individual Uranus, as in the myths, but the heavens were split up into innumerable rings. The word *ouranos* is ambiguous, and may mean (a) the sky as a whole, or (b) a part of the sky, or (c) the cosmos as a whole. Perhaps Anaximander meant it in sense (b), but was taken by Theophrastus and others to mean it in sense (c), by analogy with the Atomists. There is an alternative explanation of the evidence, as meaning that our world, although unique in its time, is born, dies, and is reborn in an infinite succession. This makes a better fit with the rest of the Milesian world picture

[14] Again, Burkert's argument (*ibid.*) that Anaximander took important elements of his ideas from Persian sources is very persuasive. A different view, that the *Apeiron* is not the container of the world but is just the infinite succession of generation and destruction of worlds and their contents, is argued by Elizabeth Asmis in *Journal of the History of Philosophy* 19 (1981).

than Atomistic innumerable worlds, and is more plausible on that ground; as we have seen, there is some evidence of a cyclical theory in the reports of Anaximander. But the reports are remarkably ambiguous and tentative, and the whole matter must remain in doubt.[15]

[15] See Kirk's discussion of this problem in Furley and Allen, *Presocratic Philosophy*, vol. 1, pp. 335–40.

4 Two philosophical critics: Heraclitus and Parmenides

4.1 *The value of* logos

The course of the history of philosophy and science was utterly changed by these two men. Many would claim, in fact, that philosophy in the modern sense began with Parmenides – or at least that a historian of philosophy today must begin with Parmenides. Earlier writers are now represented by a few isolated phrases or sentences of their own words and some garbled second-hand reports; but by great good fortune – and the far-sighted wisdom of Simplicius, who understood the historical importance of being able to quote the original text, a thousand years after it was written – a substantial portion of Parmenides' argument survives, in his own poetic words. It is undeniably an argument and undeniably philosophical; and its importance can hardly be exaggerated. Often in history, and especially in the history of science, although each progressive step is credited to the account of a single person or group, it can be seen that if the advance had not been made then, it would inevitably have been made soon by someone else. In the case of Parmenides, this is not true. There is nothing quite like his argument: it seems startlingly original.[1]

Nevertheless, there is a point in including Heraclitus with Parmenides in the same chapter, although they were separated physically from each other by the whole of the Greek world – Heraclitus lived in Ephesus, Parmenides in the South Italian town of Elea – and they wrote in different genres. Whether they knew each other's work, or whether Parmenides at least knew of Heraclitus' work, as is usually believed – these are questions that need not delay us at present. All that we need to stress are the following points of similarity. Both castigated earlier thinkers for their uncritical use of words in picking out entities in the physical universe. Both claimed that reflection on *logos* (connected, rational speech), and examination of the relation between *logos* and the universe, are the essential keys to understanding. Both argued for the conclusion that in some sense all is one,

[1] Contrast West, *Early Greek Philosophy*, pp. 218–26. The mystical and oriental passages quoted by West as 'reminiscent of Parmenides' seem to me on the contrary to highlight the originality of Parmenides' argument. The nature and value of this argument are excellently discussed by Owen, *CQ* 10 (1960).

and the apparent separation of things from each other that is brought about by giving them different names is illusory.[2]

Two characteristic quotations may serve to illustrate this.

Teacher of most men is Hesiod. They are sure he knows most – he who did not know day and night! For they are one. (Heraclitus fr. 57)

> Learn from this point mortal beliefs,
> listening to the deceitful order of my words.
> For they [mortals] have set up two forms in their minds for naming,
> not one of which should be named – in this they have gone astray;[3]
> and they separated the opposites in body, and put signs upon them,
> separate from each other: here, bright fire of flame,
> delicate, very light, everywhere the same as itself,
> not the same as the other; and that other, by itself,
> the opposites [of these], night, dark, a dense and heavy form.
> (Parmenides, fr. 8, lines 51–9)

Hesiod had spoken of Day and Night as two separate persons, and given each of them a sort of parentage. 'Dark-robed Night' was a daughter of Chaos, and a sister of Erebus, himself little more than a male version of Night. Erebus and Night then mated, and Night gave birth to Aether and Day. Night and Day live in the same house, but are never at home together: each in turn goes out to roam over the earth, and they greet each other at the gate as they pass. The point of Heraclitus' criticism, to put it simply and crudely, is this. Although Hesiod's picture shows that Day and Night are closely related, as mother and daughter, or housemates, it still gives them too much individuality. Day comes back home in the evening, and Night then goes out – but in principle, if they are two autonomous persons, each of them might do something else. Day might, perhaps, forget the route home, and Night might come out to look for her. Heraclitus claims in effect that this is conceptual chaos: Day is not the daughter of Night, but her contradictory. The sentence 'We have now both Night and Day' is not just one that has never been true because they never leave home together: it is logically impossible that it should be true. Heraclitus expresses his denial that Night and Day are two autonomous individuals in the strongest terms: 'they are *one*.' This is a highly paradoxical expression; if they are opposites, how can they be one? Perhaps the underlying idea is that they form just one polarity (Day is the polar opposite of

[2] The innovation made by Heraclitus and Parmenides is analyzed in a very illuminating article by Mourelatos, 'Heraclitus, Parmenides, and the Naive Metaphysics of Things,' in Lee, Mourelatos, and Rorty, *Exegesis and Argument*.

[3] This translation of lines 53–4 is defended in my article 'Notes on Parmenides,' in Lee, Mourelatos, and Rorty, *Exegesis and Argument*, to be reprinted in my *Cosmic Problems*. On the significance of Parmenides' discussion of 'mortal beliefs,' see below.

Night and of nothing else), or that they follow each other in a single uninterrupted temporal sequence. Parmenides also rejects the pluralism involved in the uncritical use of opposite terms. He claims that there is a fundamental error in the thinking of 'mortals' (it is a goddess who speaks these lines in the poem) in that they use words to separate two opposite forms, fire (or light) and night. But whereas Heraclitus asserts and hints, Parmenides offers an argument, as we shall see shortly.[4]

4.2 *Heraclitus on change*

Many of the surviving quotations from Heraclitus' book can be given a context and a connected meaning if we think of them as directed against giving too much autonomy to individual beings in the world.

> The same . . . : living and dead, and the waking and the sleeping, and young and old. For these transposed are those, and those transposed again are these. (fr. 88)

> The god: day and night, winter and summer, war and peace, satiety and hunger. (fr. 67)

More positively, Heraclitus expresses his view of the relatedness of things by speaking of the tension of a bow or a lyre.

> They do not understand how it agrees with itself while differing: a back-stretched connection, as of the bow and the lyre. (fr. 51)

The essence of these instruments is an opposite pull on the string: the frame of the lyre, and the two ends of the bow when it is strung, keep the string stretched by pulling its two ends simultaneously in opposite directions. The fragment as preserved does not spell out the message, but we can interpret it as an invitation to view the physical world as a set of bows or lyres, stretching the string between day and night, summer and winter, the hot and the cold, the dry and the wet, the living and the dead.[5]

Whereas the bow and lyre present an image of static tension, two other metaphors in Heraclitus suggest the constant movement of the physical world:

> As they step into the same rivers, other and other waters flow upon them. (fr. 12)

[4] The origin of Day and Night is in Hesiod, *Theogony* 123ff. (quoted above, p. 18). For the home of Day and Night, see *Theogony* 746–57.

[5] This reading of fr. 51 can be sustained whether the phrase used by Heraclitus was παλίντροπος ἁρμονίη or παλίντονος ἁρμονίη. I cannot agree with the latest interpretation I have seen (Snyder, *Phronesis* 29 (1984)), that the imagery is visual, playing on the fact that the bow and the lyre are shaped like the arc of a circle, and the ends of their frames point in opposite directions around the circle.

In other words, what our language picks out as one and the same river (we give it a name) is a body of water that is never the same from one minute to another. This is the image that was regarded as the essence of Heraclitus' philosophy by some later writers. 'All things are in flux and nothing stays': that is how Plato sums it up (*Cratylus* 402). Following Plato, later philosophers give the proposition a skeptical interpretation: since all the objects in the physical world are in perpetual flux, it is impossible to get any knowledge of them. It seems unlikely, however, that Heraclitus meant it as an invitation to skepticism. There is nothing in the extant fragments that makes the skeptical point explicitly, and much that implies the contrary. Heraclitus' message appears to be rather that *logos* – rational human language – can convey understanding of the physical world, so long as we remember that our words do not pick out isolated and static individuals. We may still speak of day and night, and give our river a name, provided that we understand the relatedness of day and night, and the flux of the river.

Heraclitus' second metaphor for the changing world is almost equally famous:

> This ordering (*kosmos*) no god nor man has made, but it ever was and is and will be: fire everliving, kindled in measures and in measures going out. (fr. 30)

But is this correctly described as metaphor? Some of the ancient doxographers supposed that fire in Heraclitus' cosmology was the basic substance from which the world was made, on the same level as Thales' water, Anaximander's Boundless, or Anaximenes' air. This is quite unlikely to be right. Milesian cosmogony, with its single *archê* producing, at some particular moment, the opposites of the physical world, is basically inconsistent with Heraclitus' conception of perpetual interchange. Moreover, fire is plainly not a substance of just the same kind as water or air: it has no determinate physical dimensions and it characteristically causes change in other things, rather than taking on different properties itself. In one aspect, it destroys: a green forest full of wild life is turned to dead, black ashes. In another, as heat, it is a cause of life: the warmth of the sun brings vegetation into new life in the spring, and the warmth of the body is at least a necessary condition of life in animals. But here it is not the fire itself that changes from the living to the dead or vice versa. Even when fire was later elevated into the original and basic substance of the universe by the Stoics, the move was possible only because of their peculiar brand of materialism (we shall revisit this doctrine in volume 2). The ancient doxographers were wrong, then, to take Heraclitus' fire as the original material substance of the world.[6]

On the other hand, Heraclitus can hardly have brought fire into his cosmology purely as a metaphor. The fire imagery differs from the river imagery in this respect. Literally, the cosmos is not a river or a set of rivers, complete with a source, banks, mouth, fords, and ferries. But in a much more literal sense it might reasonably be called, not fire, but *a* fire: it is a system in which material changes are brought about and continually maintained by the application of heat. Just as the Milesians invoked what they saw as the material basis of life – moisture and breath – so Heraclitus appeals to the sustaining cause of the processes of life. A measured amount of heat keeps the processes of growth and reproduction going. The expression 'kindled in measures and in measures going out' in fragment 30 makes best sense if we take it to refer both to time and intensity. The alternation of the seasons is the obvious paradigm for the temporal sequence of growth, decay, and then new growth. But at any time the fire must not be wholly quenched or too vigorously kindled: both excessive heat and excessive cold bring life to an end. (In the physiology of Galen and Aristotle it is the role of both the pulse and respiration to maintain a moderate heat.) So Heraclitus' insistence on 'measures' must surely be interpreted as correcting the *prima facie* image of 'destructive fire' (πῦρ ἀΐδηλον in Homer's formula).

We may suppose that Heraclitus applied this idea to what we call inanimate matter as well as to living forms. Fire is the agent of increased liveliness in matter: it transforms solid, unmoving matter into pliable stuff by melting it, and excites liquids into volatile vapours.[7]

For souls (*sic*!) it is death to become water, for water it is death to become earth; out of earth water arises, out of water soul. (fr. 36)

All chemistry is biochemistry for Heraclitus. He uses the image of fire in another context too:

All things are exchanged for fire, and fire for all things, as goods for gold and gold for goods. (fr. 90)

The first half of this might suggest, if anything in the fragments does, a

[6] Aristotle, *De caelo* I. 10, 279b12, appears to attribute to Heraclitus the view later held by the Stoics, that the cosmos is periodically consumed by fire in a universal conflagration (*ekpyrosis*), and then reborn from the fire. There is no general agreement among scholars about this. I have followed the line taken by Kirk, *Heraclitus*, pp. 307–38, in common with most recent writers about Heraclitus. Kahn revives the thesis of an *ekpyrosis* in Heraclitus in *The Art and Thought of Heraclitus*: for contrary arguments see Inwood, *Ancient Philosophy* 4 (1984), and Marcovich, *Gnomon* 54 (1982).

[7] See Fränkel, *American Journal of Philology* 59 (1938).

Milesian pattern of material growth out of and back into an original stuff. But the simile of money shows that this is wrong: gold does not turn into goods when it buys them. If we put the simile of money together with the notion of fire as the agent of change in matter, we seem to arrive at the following picture. To acquire new 'goods' in the form of matter in a livelier state, the cosmos must use up (spend) a certain quantity of fire to heat it up; this fire is restored when matter reverts to a more inert state. This is precisely the quenching in measures and kindling in measures referred to in fr. 30. It is not that fire itself is transformed into anything else; liquid (say) is transformed into vapour by the expenditure (quenching) of a portion of fire.[8]

If this is the right way of looking at Heraclitus' doctrine of fire, then we have an early suggestion of the idea of an efficient cause in nature. Compared with the post-Parmenidean theories of Anaxagoras and Empedocles, which we shall examine in later chapters, it is only a hint. They postulate efficient causes (Mind, Love, and Strife) conceptually isolated from the matter that they cause to change; the function of these entities in their systems is precisely to cause change. Fire in Heraclitus' theory has no such clear position. Being 'quenched' and 'kindled,' it is itself an ingredient in change, and it is itself one of the material constituents of physical bodies. All the same, if the increased 'liveliness' of matter is brought about through the expenditure of fire in his theory, he was plainly on the way towards the notion of a cause. Anaximander had likened the processes in the physical world to the social system of justice; and, so far as one can tell, he stayed on the level of metaphor. Heraclitus' fire functions on the same level, as a metaphor for physical change – but not only on that level. Insofar as it also functions literally in his account of physical change, it foreshadows the notion of a causal agent.

4.3 *The challenge of Parmenides*

The critical argument is more sharply distinguishable in Parmenides than in Heraclitus. It starts from a profound philosophical doubt about negation, or, as Parmenides construes it, about talking about what *is not*, or nothing.[9]

In ordinary talk about the world, we necessarily talk not only about

[8] I owe this interpretation to an unpublished paper by W. D. Furley. Something similar can be found in Wiggins, 'Flux, Fire, and Material Persistence,' in Schofield and Nussbaum, *Language and Logos*.

[9] See P. L. Heath's article on 'Nothing' in Edwards, *Encyclopedia of Philosophy*, vol. v: 'Nothing is an awe-inspiring yet essentially undigested concept, highly esteemed by writers of a mystical or existentialist tendency, but by most others regarded with anxiety, nausea, or panic. Nobody seems to know how to deal with it (*he* would, of course)...'

what there *is*, but also about what there *is not*. In particular, any description of a change must necessarily involve a mention of what there *is not*, or at any rate of what there *is not* here and now. If we say that A has changed into B, then we say that A *is not*, or there *is not* A, here and now. If there were still A here and now, then it would not have changed. So all the cosmological theories that made use of the notion of one substance changing into others necessarily involved 'what *is not*.' But Parmenides objected to this. What *is not*, is nothing; and to talk about nothing is to talk vacuously; one cannot grasp, or recognize, or point out, nothing. Any theory that necessarily involves talking about what *is not* is therefore useless.

This bleak outline needs to be filled out with the help of a translation of some lines from Parmenides' poem. It is difficult to translate into English, especially because of some differences in idiom concerning the verb 'to be.' In Greek the same verb ἔστι can be used independently, with no complement, to say of a subject 'X exists' or 'there is X,' as well as in predicative statements like 'this is a book,' or 'this is yellow.' In spite of the artificiality in English, it seems best to translate ἔστι in all its uses in Parmenides by 'is' (using 'be' and its derivatives when necessary, of course), but in the lines that follow, 'is' when italicized means exists.[10]

Parmenides puts his argument into the mouth of an unnamed goddess, who says to him:

> Come then, I will tell you – listen to my story and cherish it well –
> what are the only ways of inquiry for thinking:
> one, that it *is* and cannot *not be* –
> this is the way of Persuasion, for she attends upon Truth;
> the other, that it *is not*, and necessarily must *not be* –
> that, I tell you, is a way void of intelligence,
> for you cannot recognize what is not (that is not to be accomplished)
> nor utter it. (fr. 2)

[10] The behavior of the verb 'to be' in Greek is analyzed very fully in the works of Charles Kahn, especially in his book *The Verb 'Be' in Ancient Greek*, and with special reference to Parmenides in *Review of Metaphysics* 22 (1969). Two essential articles on Parmenides' argument are Owen, *CQ* 10 (1960), and Furth, *Journal of the History of Philosophy* 6 (1968).

Alexander Mourelatos makes a case for an interestingly different interpretation of the argument in his book *The Route of Parmenides*. I have set out some reasons for disagreeing with him in my article cited in note 3. His position is further defended, in a somewhat modified form: 'Determinacy and Indeterminacy, Being and Non-being in the Fragments of Parmenides,' in Shiner and King-Farlow, *New Essays on Plato and the Pre-Socratics*.

In the book, Appendix II, Mourelatos discusses six possible interpretations of the 'bare ἔστι' as in frs. 2.3 and 2.5, and suggests that '– is –' is the nearest equivalent. In the later paper, he concedes that it 'carries a semantic component of the existential nuance' (p. 47). I believe this is a change in his position, rather than an elucidation; but he still disagrees with me in seeing the indeterminacy of what *is not* as the reason for dismissing it from discourse, rather than its unavailability as a referent.

> For the same thing is for thinking and for being. (fr. 3)
>
> What can be spoken and thought of must *be*. For it can *be* – but Nothing cannot. (fr. 6.1–2)

If we are to inquire into something, or seek for it (the nature of the cosmos, for example, or its *archê* – but initially at least Parmenides leaves the object quite open), then we can envisage at once two possibilities concerning the object of our inquiry: that there *is* and *must be* such an object, and that there *is not* and *cannot be* such an object. But the second of these ways can be dismissed almost as soon as it is articulated. *Nothing* (a thing that *is not*) cannot be; hence it cannot be recognized or spoken of; hence it cannot be an object of inquiry.[11] Parmenides then sets out an argument to show by elimination that only the first way remains: he calls it the Way of Persuasion, but it is usually referred to as the Way of Truth.

What then can be said about the object of inquiry, if we take the Way of Truth? Parmenides discusses the properties that it must have in the long fragment 8. First, what *is* is ungenerated and indestructible. The only thing that it could be generated from or destroyed into is what *is not*, and so we could not describe its generation or destruction without using this now forbidden notion. Secondly, it is one and indivisible, because the only thing that could divide it is something other than itself, and that could only be what *is not*.[12] Thirdly, it is motionless and unchanging, since there is nothing other than itself into which it could move or change. Fourthly, it is complete, or perfect, without defect: or as he expresses it, 'like the mass of a well rounded ball, equally balanced from the center everywhere,' since it contains no element of what *is not*, which alone might constitute a variation in its texture.[13]

It follows from this argument, insofar as it is successful, that the world as presented to our senses is not a possible object of inquiry. The senses report that things in the world come into being and perish, that they are distinguishable from one another by differences in perceptible quality,

[11] The argument is complicated by the modal verbs, 'cannot' and 'must.' Jonathan Barnes has set out a clever and plausible analysis of the structure of this argument in his recent book *The Presocratic Philosophers*, pp. 163–5.

[12] Barnes, *ibid.*, and in his article, *Archiv für Geschichte der Philosophie* 61 (1979), has raised doubts about whether Parmenides produced any argument to show that what *is* is all one. I believe it is to be found in fr. 8.22–5, where he aims to show that what *is* is undivided and continuous. Barnes thinks that this shows only that if a thing exists then it is undivided and continuous. It seems to me to work just as well if we take the expression 'what is' to mean 'all that is'; in that case the conclusion means that there is just one thing in existence.

[13] I have more to say about the fourth of these properties of what *is*: see below, pp. 54–7.

and that they move around in space. But Parmenides' argument purports to show that none of these can be properties of what *is*: hence there must be an unbridgeable gap between what *is* and perception, between being and seeming.

After discussing the four properties of what *is*, Parmenides' goddess continues with the lines we quoted at the beginning of this chapter:

> Here I end my trustworthy discourse about Truth:
> learn, from this point, mortal beliefs,
> listening to the deceitful order of my words.
> For they [mortals] have set up two forms in their minds for naming,
> not one of which should be named – in this they have gone astray;
> and they separated the opposites in body, and put signs upon them,
> separate from each other: here, bright fire of flame,
> delicate, very light, everywhere the same as itself,
> not the same as the other; and that other, by itself,
> the opposites (of these), night, dark, a dense and heavy form. (fr. 8.50–9)

This is a much disputed passage and there is still no agreement among scholars on what exactly it means, or even on how it should be translated.[14] It is clear that the Way of Truth has now ended, and that the four properties are all that the goddess finds to say about what *is*. Since one of the four properties is that of being undivided, there cannot be two forms, and since another is that of being unchanging, there cannot be a succession of Day and Night. Yet the ordinary speech of human beings uses these two opposed names, and uses them as though they both named something real, something that *is*. From this primary pair of opposites, they go on to others, such as hot and cold, rare and dense, and so on, and so generate the familiar descriptions of the sensible world with its various perceived properties. Parmenides' poem goes on (unfortunately now in a very fragmentary state) to set out an astronomy, zoology, embryology – the limits are uncertain. The poem thus continues along the third 'way of inquiry' mentioned – and rejected – in some earlier lines, immediately following the demolition of the Way of Not-Being:

> I hold you back, too, from this way, on which mortals, knowing nothing,
> wander, two-headed; for helplessness in their breasts
> steers their wandering mind. They move
> like men who are deaf and blind, dazed, undiscerning hordes,
> by whom to be and not to be are deemed the same
> and not the same, and the path of all is backward turning. (fr. 6.3–8)

What are we to make of this third way, described as the Way of Mortal

[14] For discussion, see my 'Notes on Parmenides,' and most recently Leonard Woodbury's paper delivered to the Society for Ancient Greek Philosophy in Toronto, December 1984.

Doxa (belief, opinion, seeming)? Whereas the previous two Ways are labeled with the expressions 'it *is* and cannot *not be*' and 'it *is not*, and must *not be*,' the third is labeled – although not very clearly – 'to be and not to be.' Its concepts, such as daylight and night, or light and heavy, involve both being and not being. Parmenides' claim is something of this sort: when men make distinctions between such opposites, they do so only by treating what *is not* as if it were something that *is*. They 'set up the forms' of daylight and night for naming: but they cannot pick out either of them without the other. Yet the non-being of the one is entailed by the being of the other: if there *is* daylight, then *ipso facto* there *is not* night. The significance of its being daylight depends on the denial of its being night. There cannot be such a thing as daylight unless there *is* such a thing as night, the *non-being* of which is entailed by daylight.

The cost of Parmenides' claim that 'you could not recognize what *is not*, nor speak of it' is great. The perceptible world, characterized by incompatible properties, is not a proper object of inquiry: it can be described only in 'a deceiving order (*kosmos*) of words.' Why, then, describe it at all? Why did Parmenides write about the visible and moving world of the stars, sun, and moon, and the generation of animals? The only explicit hint in the verses that survive is given in a metaphor:

> I tell you this ordering (*diakosmos*), all fittingly disposed,
> so that mortals' thought may never drive past you. (fr. 8.60–1)

The metaphor of 'driving past' may suggest that the hearer of the Way of *Doxa* is a competitor in a race with other mortals' opinions – one who will never be defeated. But how can that be so, if the whole Way is false? Perhaps, then, the metaphor merely suggests the impossibility of defeat in argument: the hearer of the Way of *Doxa* will never be overwhelmed by any proposed cosmology, so as to believe it to be true, because he understands that there cannot *be* two opposite forms.[15]

But it is still remarkable that Parmenides' Way of *Doxa*, advertized as untrustworthy and deceiving, and as containing a radical incoherence, seems to have put forward certain cosmological doctrines that both received and deserved recognition as progressive and even path-breaking

[15] In *Phronesis* 27 (1982), Mary Margaret Mackenzie points out that *both* the Way of Truth *and* the Way of *Doxa* present paradoxes, and thus form a dilemma. The Way of Truth argues that reason must conclude that there is only one being – and hence that there is no reasoner distinguishable from the rest of being. But the way of *Doxa* shows that if we allow the existence of mortals and their power of distinguishing things by naming them, we are committed to the being of not-being. I find this attractive, because it brings out the character of Parmenides' argument as a challenge (see p. 41) – both Ways seem to be impassable, and hence we must look for another way out. (The pattern of argument is one that we shall recognize again in Zeno, chapter 8 below.)

discoveries. He asserted that the evening star is the same physical object as the morning star, so we are told, and he was the first to hold that the earth is a sphere lying at the center of the cosmic sphere (Diogenes Laertius VIII. 48 and IX.21). We have his own words (fragment 14) asserting – for the first time in the surviving records – that the moon shines with reflected light. These indications seem to point to a great advance towards the geometrical, spherical cosmology of Plato and Aristotle.[16] It is hard to believe that Parmenides did not recognize his position as making an advance. If so, we find him simultaneously claiming that his is a better cosmology than any other, and that it is nevertheless false.[17]

We shall recognize a somewhat similar pattern again in Plato. The philosopher, following up what appears to be the logic of his thought, argues himself out of existence, or at best into a shadowy and inferior class of being. Plato's separation of the world of becoming and the world of being, in the *Republic* and the *Timaeus*, seems to put him in this position: Parmenides is more unambiguously there. He could not have lived, if the Way of Truth were true: if there is no birth, then Parmenides was not born. Yet that is beside the point. The argument of the Way of Truth should be regarded not as claiming to describe how the world is, as offering an alternative cosmology, but rather as a sort of challenge.[18] 'Here we have an argument; its consequences are such and such. Now you – the hearer – must either make what you can of the consequences, or else show that there is an untrue premiss or an invalid step in the argument.' It is not irrational that the writer should then continue the conversation at another level, explicitly acknowledging that we are now adopting different premisses.

By exhibiting a gap between what could rationally and coherently be

[16] Did Parmenides learn some of these ideas from the school of Pythagoras at Croton? At some stage in their history, the Pythagoreans worked out a cosmology in which the geometry of the sphere was all-important. We shall examine it in chapter 5. The evidence about Pythagoras himself and about the early school is extremely scanty: see Guthrie, *HGP*, vol. I, chapter 4, and Burkert, *Lore and Science in Ancient Pythagoreanism*, chapter 4. The controversy about the priority of Pythagoras or Parmenides began in antiquity: see Diogenes Laertius VIII.48.

[17] See below, section 5.2, for more on Parmenides' direct contribution to cosmology.

[18] Why, then, is the argument presented as 'the Way of Persuasion (for she attends upon Truth)'? I think this interpretation is consistent with the oracular character of the goddess' statement. The hearer must puzzle out its application to the physical world. The opening lines of the poem (fr. 1) present the author in the guise of a religious initiate, receiving a communication from an unnamed goddess. It is not an evil fate, she tells him, that has brought him on the road to her house, but Right and Justice: that means that there are others to whom this communication could not rightly and justly be revealed. It needs a 'man of knowledge' (line 3) to receive it. This surely means, at the least, that the hearer must bring some gifts of interpretation to the message, before its significance for his own life and understanding can become clear.

said about *being* on the one hand, and what appeared to the senses of an observer in the physical world on the other, Parmenides set the agenda for most of Greek philosophy for the next two centuries. The struggle to close the gap, or bridge it, or (finally) to show that it did not exist, will occupy us in the following chapters.

4.4 *The effects of Parmenides' poem*

Parmenides pulled away the ladder by which previous and contemporary writers had been content to climb. For three or four generations after his work, it was a question of rigging up a makeshift alternative; the ladder was not securely replaced until the time of Plato – late in his career – and Aristotle.

In order to understand the difficulty encountered by philosophers immediately following Parmenides, especially Anaxagoras, Empedocles, and the Atomists, it may be helpful first to study briefly the defense against Parmenides' challenge that was finally worked out by Plato, in his dialogue, the *Sophist*. This may give us a clearer idea of what it was that earlier philosophers lacked – a lack that determined some of the peculiar features of their theories.

As often in the Platonic dialogues, the metaphysical discussion is embedded in an unlikely surround – a half-serious search for the right definition of the sophist. The Parmenidean problem comes to the surface during the discussion of a definition of sophistry as a kind of image-making. Rather artificially, Plato distinguishes two branches: that in which the image reproduces exactly all the features of the original as they *are*, and that in which the image reproduces the object as it *seems*. For the latter, which is called φανταστική, he cites the example of a colossal sculpture in which the upper parts are made disproportionately large so that they will *appear* proportionate to a viewer at ground level. The sophist is like such a *phantastic* sculptor in that he produces, in words, 'images' that *seem* true and yet are not.

At this point, Plato calls to mind what Parmenides had written. This notion of the sophist has described him both as saying *something* – something that seems to be true – and as saying what *is not* – i.e. what is not the case. The definition thus seems to contradict itself. 'This account [of the sophist],' says the Eleatic Stranger in Plato's dialogue, 'has had the audacity to suppose that what *is not*, is; for falsehood could not otherwise come into being. But the great Parmenides, my young friend, when we were your age, testified against that from beginning to end, in his poem and elsewhere always telling us this:

Never shall this prevail, that things that *are not*, are:
you in your inquiry must hold back your thought from this way.' (*Sophist* 237a)

Parmenides had attempted to demonstrate what *can* be said about Being without talking at all about what *is not*. In the Way of Truth, he argued that what *is* has just these properties: it is ungenerated and indestructible, a single uninterrupted whole, immovable, and complete. Plato now shows that if we rule out the concept of what *is not* our situation is actually much worse than Parmenides had claimed. For example, we cannot even mention what *is not* legitimately, because every expression such as 'what *is not*' is either singular or plural, and number is a property of what *is*, not of what *is not* (*Sophist* 238a–239a). Moreover, so far from being in a position to spell out the properties of what *is*, as Parmenides did in the Way of Truth, we cannot say anything at all about it. We cannot even give it a name: if it had a name, then there would *be* two things, what *is* and the name 'what *is*,' and this is refuted by Parmenides' argument that if what *is not* is unthinkable, what *is* must be a single, uninterrupted whole (*Sophist* 244d).

If we are to get anywhere at all, Plato says, 'we shall have to torture the argument of father Parmenides, and force the admission that what *is not* in some respect *is*, and that what *is* in a certain way *is not*' (*Sophist* 241d). The problem that has led to this intolerable impasse is that the expression 'what *is not*' appears to refer to nothing. Plato now introduces an item called 'the other' or 'the different' (*to heteron*) and shows that many of our sentences involving 'what is not' can be explained as referring to this entity, which has a perfectly uncontroversial claim to be admitted to our ontology. The sentence 'X is not Y' is to be analyzed as 'X is other than Y'; 'is not' has thus disappeared, taking the problem with it.

Since this solution seems remarkably simple and obvious, it may be useful to point out certain features of the Greek language that helped to conceal it. In the first place, as I have mentioned already, whereas English uses the verb *exist* in addition to the verb *be*, Greek uses the single verb *einai*. There are contexts in which English uses the verb *be* existentially as Greek does; this very sentence 'there are contexts . . .' is one of them. But English habitually uses different verbs in sentences like 'hybrid forms exist' and 'hybrid forms are easy to grow'; in Greek, the same verb is used in both of the equivalent sentences. Hence, although English philosophers have often discussed the problem about how we succeed in talking about what does not exist, that problem does not naturally infect predicative sentences in English as it does in Greek. Secondly, whereas English predicative sentences usually have a distinctive word order, 's is P,' this is not the case in Greek, where the verb may come first, or last, in a predicative or an existential sentence. Given the distinction between existential and predicative sentences, it is sometimes easy for us to classify Greek sentences involving *esti* into one box or the other; but it is often difficult and sometimes impossible. The distinction was slow to emerge in Greek philosophy; and in some sentences the context shows that both existence and predication are signified by the verb. Existence and predication are 'fused' into one.[19]

To avoid using Greek we can represent Greek idiom symbolically thus. Let 'I' represent 'exist,' 'exists,' 'is,' 'are,' 'there is,' 'there are' (sometimes 'here is' and 'here are'); and 'N' the negative of these. Then we have the following translations:

(1) Hybrids exist = I (hybrids)
(2) Centaurs do not exist = N (centaurs)
(3) Hybrids are easy to grow = I (hybrids, easy to grow)
(4) Hybrids are not cheap = N (hybrids, cheap)
(5) This rose is yellow = I (this rose, yellow)
(6) This rose is not red = N (this rose, red)

The contents of all the brackets following 'N' can be thought of as members of the class of things that *are not* (or the class of what *there is not*). And that is where Parmenides found his problem: in talking about what there *is not*, one seems to be talking about nothing.

Plato now proposes, we might say, to introduce the symbol 'O', representing '— is other than —,' and claims that sentences (4) and (6) should be rewitten:

(4′) (hybrids) O (cheap)
(6′) (this rose) O (red)

There is now no problem of reference in these sentences, although there

[19] See Kahn, *The Verb 'Be'*, pp. 164–7, 271–7, and elsewhere. Also his article in *Archiv für Geschichte der Philosophie* 58 (1976), and Furth, *Journal of the History of Philosophy* 6 (1968).

A clear example is quoted by Kahn, *The Verb 'Be'*, p. 165: μέσσῳ δ'ἐν σκοπέλῳ ἐστὶ σπέος ἠεροειδές, 'In the middle of the crag is a dim cave' (*Od.* 12.80):

Jonathan Barnes argues (*The Presocratic Philosophers*, vol. I, pp. 159–61) that nothing but the existential use of *esti* is to be found in Parmenides. To my mind, his attack breaks down on the Eleatic argument against change. If everything is to be in strictly existential terms, with subjects of the verb 'to be' confined to the kind of thing that is normally said to exist or not to exist, then the argument against qualitative change has to depend on the abstraction of qualities as existing subjects. Can Thrasymachus blush? No, because that entails the non-existence of Thrasymachus' pallor, which now exists. It seems to me more likely that, if questioned on this point, Parmenides would claim that 'Thrasymachus-pale *is not*' is the objectionable thought.

may still be one in sentence (2). We can make distinctions within the realm of what *is* without irrationality. Given that we can make distinctions, it is possible now to recognize and talk rationally about plural beings, and about differences of quality, change in time, and motion in place. We can say 'this object is not that,' 'this is not green,' 'this is not as it was,' 'this is not where it was,' recognizing that these are all, in Plato's language, derivatives of *the other*. Whether Plato himself ever used this argument to rehabilitate the possibility of rational discourse about the *physical* world is a controversial question, to be discussed in volume 2. In the *Sophist*, his concern is to make distinctions between concepts, or Forms, and to discuss the relations that hold between them. But Aristotle handled the Eleatic problem of negation and reference in a similar way to Plato, substituting *privation* for *otherness*, and it was this that enabled him to discuss plurality and change in the physical world without feeling any metaphysical embarrassment when he re-read Parmenides.

But the point of this excursion into later developments was to illuminate the struggles of Anaxagoras and Empedocles, still entangled in the Parmenidean net. Lacking the device of reinterpreting 'what *is not*' as 'the other,' they attempted to do physics without allowing 'what *is not*' any place in it at all. 'What *is* cannot *not be*,' said Anaxagoras (fr. 3).

> Coming-to-be and perishing are customarily believed in incorrectly by the Greeks, since nothing comes-to-be or perishes, but rather it is mingled together out of things that *are*, and is separated again. Thus they would be correct to call coming-to-be 'being mingled together,' and perishing 'being separated.' (fr. 17)

Empedocles wrote:

> Something else I will tell you: there is growth (*physis*) of no one
> of all mortal things, nor end in baneful death;
> but only mingling, and separating of things mingled,
> is the case; and 'growth' is men's name for these. (fr. 8)

> From what *is not* at all, it is impossible that something should come to be,
> and that what *is* should perish is unmanageable and unintelligible,
> for it will always *be* wherever one may keep pushing it. (fr. 12)

It was not just *creatio ex nihilo* that they wanted to deny: no one in the early history of Greek philosophy ever asserted that an existing thing, such as the world, came into being out of 'the non-existent' in the sense of nothing at all. They used the verb *be* in a manner that fuses the predicative with the existential, as we have just explained. What troubled them was the notion that we should have to suppose that a state describable in the symbolism we introduced on p. 44 as 'I (x)' should be derived from one describable as 'N (x),' with x the same in both. In other words, they denied that a thing could come into being from what *is not that thing*. 'How could hair come from not hair,' Anaxagoras asked, 'or flesh from not flesh?' (fr. 10).

Although they thus accepted the Parmenidean ban on coming-to-be, Anaxagoras, Empedocles, and the Atomists all made one move away from Parmenides in common with each other. That was to recognize the possibility of making distinctions between one part of what *is* and another: the possibility of plurality, in other words. Parmenides had written:

> And it is not divisible, since it *is* all alike
> nor is there any more of it in one place, which might prevent it from holding together,
> or less of it [anywhere]; but all is full of what is.
> Therefore all holds together, for what *is* is next to what *is*. (fr. 8.22–5)

> Look just the same at things distant as being steadily present to the mind;
> for you will not cut off what *is* from clinging to what *is*. (fr. 4.1–2)

Since what *is* admits of no variation whatever, it has no internal distinctions at all: the mind cannot find any ground on which to distinguish this part from that. This proposition of Parmenides was denied by his pluralist opponents. They asserted that there *are* different kinds of beings, distinguished from each other by their properties but all equally *beings*. It is not clear whether they advanced any argument against Parmenides in defense of this assertion, or whether they just dogmatically rejected Parmenides' incredibly austere ontology. Parmenides' monism was based on the rejection of dualism. To put it as simply as possible, he had argued that to distinguish one thing from another is to say 'I(x) and N(y)' ('here is x and not y') – but this is to say that y is a non-being, which is unintelligible. The Pluralists seem to have claimed that 'I(x)' can stand by itself and does not need 'N(y)' to make it intelligible. We can identify the objects that can replace x in 'I(x)' not negatively, but positively – primarily by the use of our senses. Reports of the senses, such as 'here is light,' 'here is green,' 'here is something hard,' and so on, are each individually intelligible. So whereas Parmenides claimed that what *is* is everywhere the same, and so the mind can find no distinctions in it, Empedocles writes:

> But come, consider by every means, in whatever way each thing is clear,
> holding no sight more in trust than hearing,
> nor clamorous hearing more than the evidence of the tongue;

> nor from any other bodily sense, by which there is a path for understanding,
> hold back your trust: but understand, in whatever way each thing is clear.
> (fr. 3)

If the Pluralists claimed that beings can be distinguished from each other without lapsing into nonsense, it may be asked how their handling of Parmenides' problem differed from the solution proposed in Plato's *Sophist*. In effect, it might be said, they had already introduced into their ontology the concept of the Other; all that Plato did was to make the move explicit and set out its stages.

But this misses a crucial difference. Plato showed that there is a use of 'is not' that is free from the objections of Parmenides. The Pluralists were still embarrassed about the use of 'is not.' Plato's solution showed the way to a coherent account of coming-to-be and passing away; the Pluralists were still in a position where they had to do without them.

Once it has been conceded that a plurality of beings is possible, change in the physical world can be accounted for without lapsing into unintelligibility. Beings can simply change places with each other – that is all the apparatus that is needed, and it is all that is used by the Pluralists who came after Parmenides. All physical change, in their theories, is reduced to *rearrangement*. Unchanging beings shift from one place to another. Thus, nothing comes to be out of what is not that thing, and Parmenides' ban is maintained.

So far we have mentioned what the Pluralists have in common. What differentiates them from each other is the range of things they admit to the class of beings. In the case of Empedocles and the Atomists, the evidence leaves no room for doubt or controversy about their theory in this respect: Empedocles' ontology contains Earth, Water, Air, and Fire, together with two motive forces called Love and Strife; the Atomists' is confined to Atoms moving in a Void. All physical objects in the world, in these theories, can be *reduced* to quantities of these irreducible beings, and all physical change is a rearrangement of these unchanging elements.

Anaxagoras' ontology is more obscure and controversial. In chapter 6, I shall put forward my own view of it, which is that he attempted the task of doing without the idea of reduction altogether. An example may make the difference clear. One of the phenomena the philosopher of nature is called on to explain is nutrition, or as Anaxagoras puts it, 'how could hair come from not-hair, or flesh from not-flesh?' We consume beef, lettuce, wine...; we grow, among other things, hair. Empedocles' explanation takes the form of *reducing* beef, lettuce, wine, and hair to the same elements, namely Earth, Water, Air, and Fire. Beef is 'really' a quantity of these four in combination: the same elements, in a different combination, constitute hair. What appears to be a case of 'hair from not-hair' is reduced to '*this* combination of earth, water, air, and fire, from *that* combination of earth, water, air, and fire,' and the problem vanishes. The Atomists' solution is of the same type, substituting atoms of such and such shapes and sizes for the four Empedoclean elements. Anaxagoras, on the other hand, does not *reduce* the beef and the hair to common elements: he asserts instead that the one contains the other. If one eats beef and grows hair, there was *hair* in the beef. For every change, if an X changes into A, B, C, . . . , then there was A, B, C, . . . in the X. And this goes back, necessarily, to the beginning of the world. Anaxagoras began his world – and his book – thus: 'All things were together.' That is to say, *everything there is* in the world has been there from the beginning; all that changes is the way the ingredients are combined.

In later chapters, after some thoughts about those who gave more attention to the structure of the cosmos than to its material constituents, we shall look more closely at each of these three pluralist theories in turn. According to my assessment, Anaxagoras' 'pantontology,' if we may so call his theory that all the things that *are* are basic and original ingredients of the universe, constitutes a more primitive and less fruitful response to Parmenides than the other pluralist theories. Empedocles' invention of elements, and the Atomists' stripping down of the elements to an irreducible minimum, were of much greater historical significance. I shall, therefore, discuss Anaxagoras before Empedocles, although it is not agreed which of them preceded the other in time.

The invention of elements is indeed a notable step in the progress of natural science. It is particularly striking that the step was taken as a response, not to any observations of natural processes, but to a metaphysical argument of the most abstract kind.

5 Pythagoras, Parmenides, and later cosmology

The beginning of the fifth century B.C. – the time when Greek civilization was menaced by the expanding might of the Persian Empire, and managed to save itself from being overwhelmed – is a period of some confusion for the literary historian. The materials he has to work with consist of the Victory Odes of Pindar, some other specimens of lyric poetry, and a few of Aeschylus' plays. In philosophy, we have a chancy collection of the fragments of Heraclitus, Parmenides, Anaxagoras, and Empedocles. Hardly anything else is preserved in sufficient quantity to allow a critical view to be taken of it, or even to provide background against which we can locate the works that have survived. It is possible that our histories do some injustice, by concentrating on the known names and works, to others whose efforts were at the time of equal importance. But we have no choice: we can only endeavor to make the most of the available evidence. Some have tried to supply the background that is missing from the Greek inheritance by looking at other cultures, especially those of the Near East.[1] It is, of course, right to make the effort, and for the period when the transmission was not by written texts but oral, and when myth dominated the poetic scene, the comparisons have proved invaluable.[2] In my opinion they have been much less successful and illuminating for the period when, with the growth of literacy, philosophy and the exact sciences were emerging from the mythical background. As we observed in the last chapter, no parallels have been found in Near Eastern myth for the argument of Parmenides.

The confusion and incompleteness of the evidence for this early period is particularly exasperating with regard to the most mysterious figure in the history of Greek philosophy, Pythagoras. That there was such a person is certain. He was probably known to Xenophanes, who refers to the doctrine of transmigration associated with him (fr. 7). He was named by Heraclitus in two extant fragments (40 and 129); so his lifetime cannot have been much later than Anaximander and Anaximenes, and he may have known their work. He was born in the Aegean island of Samos (where an attractive town has recently been named after him) in the first half of the sixth century, and migrated to the Greek colony of Croton in

[1] See most recently West, *Early Greek Philosophy*.
[2] Some of them are referred to in chapter 3.

South Italy, reportedly to escape from the tyranny of the Samian ruler Polycrates, in the 530s or thereabouts. Very soon, apparently, he became a person of political influence in his adopted city. In the fifth and fourth centuries the cities of Croton and Metapontum, especially, are named in connection with 'Pythagoreans,' who held political power in these cities from time to time. Although some individual names are known, the Pythagoreans are more frequently referred to in the plural. This is plainly no accident. More than any other philosopher before Plato, Pythagoras was associated with a school, or a sect. Pythagoreans established a boundary between themselves and others, not only by subscribing to certain doctrines, but also by adopting a communal way of life that included strict and elaborate taboos: vegetarianism was a major feature of this. Pythagoreans believed in the kinship of all living things – an idea reflected in the most famous of their doctrines, the transmigration of souls from one kind of living creature to another.

That so powerful a figure was unknown to Parmenides is extremely implausible. Parmenides' home town of Elea is not much more than a hundred miles from either Croton or Metapontum. The extent of Pythagorean influence on Parmenides – and vice versa – has often been discussed, and very different views have been taken of it: some have gone so far as to believe that Parmenides was a member of the Pythagorean community, and that the second part of his poem was a kind of summary of Pythagorean views.[3] As we shall see, the last clause, at least, is unlikely to be true. But there are nevertheless certain aspects of Parmenides' philosophy and of Pythagoreanism that have enough in common to warrant talking about them together.

This will enable us to develop a point of great importance to this history: Parmenides' poem had a seminal role in the emergence of both types of cosmology in Greece. In chapters 6 to 11 we shall examine the course of the arguments that led from Parmenides' Way of Truth to the materialist theories of Anaxagoras, Empedocles, and Democritus. But in the present chapter we shall also see some first traces of the geometrical cosmology of Plato's *Timaeus*, and of Aristotle's centrifocal theory of motion. More significantly, we shall recognize the first notes of the theme that becomes the *leitmotif* of Aristotle's criticism of the materialists, as I shall argue in chapter 13: the vital importance of form and structure, as opposed to material components, for our understanding of what makes things as they are.

[3] See Burnet, *Early Greek Philosophy*.

5.1 *Mathematics and cosmology*

Aristotle, as always one of the most important sources of information, attributes to the Pythagoreans certain ideas about mathematics, closely akin to some of Plato's ideas and therefore hard to disentangle from them, and a cosmological system. The name 'Pythagoras' Theorem' shows the strength of the tradition that connects the founder himself with mathematics; on the west front of Chartres cathedral Pythagoras sits hunched over his geometrical instruments in a niche close to Aristotle. The tradition has been doubted, and there are few scholars now who will defend the proposition that the old Samian was himself the author of a systematic science of geometry. What is clear is that mathematics had *some* history in Greece before the great developments in Plato's Academy, that the Pythagoreans had some part in this history, and that their particular contribution may well have concerned the relation of mathematical theory to cosmology.[4] So we shall first take up this latter theme.

Aristotle begins his account of the Pythagoreans in the *Metaphysics*, just before he describes the philosophy of Plato, with these words:

> Among these philosophers [*sc.* Leucippus and Democritus – 'among' presumably means no more than 'contemporary with'] and before them, the Pythagoreans, as they are called, were the first to deal with mathematics and they made advances in this field; moreover, having been brought up in it, they believed the principles of mathematics to be the principles of everything there is... Since all other things seemed in their whole nature to be modeled on numbers, and numbers to be first in all of nature, they held the elements of numbers to be the elements of everything there is. (*Metaphysics* 1.5, 985b23–986a1, excerpted)

This is plainly a highly significant statement. Aristotle has previously described theories that declared 'the principle (*archê*) and element (*stoicheion*)' of everything there is in the world to be the material out of which things emerge (see above, p.19). He referred to the single material *archê* of the Milesians and to the plural material elements in the theories of Anaxagoras and Empedocles, and just before our quotation he spoke of atoms and void as being the elements for Leucippus and Democritus. But now he records a very different idea of elements: the elements of the physical world are the elements of numbers. We must inquire what these elements were, and what were the similarities that made the Pythagoreans bring together number and the sensible world. But right from the beginning it seems obvious that their elements could hardly have been supposed to

[4] For a brief, and skeptical, view of the problem, see Heidel, *American Journal of Philology* 61 (1940). For another view, Knorr, 'On the Early History of Axiomatics,' in Hintikka, *Proceedings of the Second Conference of the International Union for the History and Philosophy of Science*, vol. 1.

function in just the same way as earth, water, air, and fire, or atoms and void.

The first principles of all number, according to this theory, were Limit and the Unlimited.[5] This must mean that any number – i.e. any cardinal number – must be thought of as a determinate piece carved out of a potentially infinite plurality: a number thus contains some element that makes it determinate and another that makes it plural. Moreover – and this is what Aristotle recognized as the characteristic Pythagorean contribution – numbers themselves, so constituted out of Limit and the Unlimited, function as the elements of physical objects: sometimes Aristotle goes so far as to say that in their view 'physical bodies are made of numbers.'[6] This is much harder to understand: Aristotle found it incomprehensible. The characteristic quality of nature, in his view, is the capacity for change, and yet mathematical objects such as numbers are changeless. Physical objects are extended in space, but numbers are not; and even if they were, how could they acquire properties such as lightness and weight? If the cosmos itself is constituted by some number or array of numbers, how can these same numbers function as the causes of the coming-to-be of particular things?[7]

It would appear that the Pythagoreans hit upon the idea – to grow into paramount importance in the fourth century – that the form and structure of physical things determines their nature to a greater extent than their material components. If number has any claim to be the essence of a physical object, it must be in the sense that some numerical formula expresses its characteristic structure. Perhaps a hint can be drawn from the technique, attributed to some Pythagoreans, of representing the structure of a thing by an array of pebbles.[8] It is a device that may be thought to have a *prima facie* resemblance to Atomism, but it is not actually Atomistic in inspiration: the pebbles are not thought of as the bricks that, when piled together in a particular way, make a house. They are not the parts into which the thing can be divided. Rather they represent certain proportions characteristic of the thing. There is a strong tradition that Pythagoras himself discovered the relations between simple numerical ratios and the three primary concordant intervals in music: the octave is produced by the ratio 1 : 2 (for example if a lyre-string is stopped halfway along its length), the fifth by 3 : 2, and the fourth by 4 : 3.[9] And this discovery may have been the inspiration for the extension of numerology to other things.

[5] Aristotle, *Metaphysics* I.5, 986a13–21. [6] *Ibid.*, XIII.8, 1083b10.

[7] For these criticisms, see *ibid.* I.8, 989b21–990a32.

[8] *Ibid.* XIV.5, 1092b10–12, with the ps.-Alexander commentary, 827.9ff.

[9] A quotation from Plato's contemporary, Xenocrates, is the earliest source for this (Porphyry, *Comm. in Ptolemaei Harmonica*, ed. Düring, p. 31.1 = Xenocrates fr. 9 Heinze). See Guthrie, *HGP*, vol. I, pp. 212–26 for an excellent discussion.

Nevertheless, although this thesis about number appears to be a statement about structure, it seems that the earlier Pythagoreans were not able to find the language to distinguish it from a thesis about matter. Physical objects and numbers had the *same* elements: they did not say that some separate matter *possessed* or *manifested* the numerical structure. Matter and structure somehow coincided. (Aristotle found a similar puzzle in Plato's *Timaeus*: see *De caelo* III.1.) This theory included a cosmogony of sorts. The first product of the basic pair of principles, Limit and the Unlimited, is the One; the One, now identified as the cosmos, 'breathes in' the Unlimited, now described as the void, and the void serves to 'delimit natures, the void being a kind of separation and determining of things next to each other.'[10] The failure to separate out the ideas of form and matter is illustrated particularly well by this doctrine. No doubt Aristotle is responsible for sharpening the point up to suit his own purposes, but we need not doubt that he found something of the sort in his sources.

5.2 *The pioneer of the centrifocal universe*

In chapter 3 I argued that the Milesian universe was a linear one: it possessed a single downward direction, like the unsophisticated picture of the ancient cosmological myths. This is in contrast with the centrifocal universe of Aristotle's cosmology, in which all directionality is focused on the central point of a sphere. It is important to be clear on the full extent of this distinction. It is true that the circle and its center figured quite largely in Milesian cosmology: Anaximander's sun, moon, and stars were contained in circular rings, and the earth lay on the line through the center of all the rings. According to one testimony (DK 12A10), the rings formed a sphere before they were separated off to what we conjecture to be their proper intervals of 9, 18, and 27 earth-diameters. But unless I am mistaken about Anaximander, his universe was still conceived in linear terms, like that of Thales and Anaximenes.[11]

His earth was flat, with a top side and a bottom side. He had not made the great leap of the imagination required to picture the sun setting over Miletus as being vertically above some other part of the earth's surface, and every part of the sky as having an equal claim to be 'up.'

There is a great deal of confusion about the first proponent of centri-

[10] Aristotle, *Physics* IV.6, 213b22–6. The expression 'breathes in the void' suggests that this theory belongs to the primitive period before Anaxagoras and Empedocles had deliberately shown that air is corporeal.

[11] See chapter 3, and my article referred to in note 11. If I am wrong, we must simply shift the origin of the centrifocal universe a few decades back in time and several hundred miles east. The distinction between the linear and centrifocal universe remains just as valid.

focality, in ancient as well as modern literature, but I am convinced that the best candidate for this Gold Medal is Parmenides:

> But since there is an outermost limit, it is perfected
> from all sides, like the mass of a well-rounded ball,
> equally balanced from the centre everywhere. For neither greater
> nor smaller must it be in one place or another.
> For neither is there Not-Being, which might stop it reaching
> its like, nor is there Being such as to be
> more than Being here, less there, since all is inviolate.
> For equal to itself from all sides, it lies uniformly in its limits. (fr. 8.42–9)

But before this section of the Way of Truth can have any claim to be accepted as the forerunner of the centrifocal universe, two highly controversial points of interpretation must be explained and defended. The first concerns the subject of the first sentence, the 'it' that is perfected, like a ball. According to some historians of philosophy, Parmenides' argument here is a highly abstract deduction of the entailments of being as such. If that is right, the language of this section must be taken figuratively: the limit, the perfection, the well-rounded ball, the equal balance, all signify that the concept of being is a unity that has no varieties of degree or quality. I prefer a somewhat less abstract reading, according to which Parmenides' subject is not the notion of being as such, but rather that which *has* being, or what *is*. Since the passage under discussion is at the end of the Way of Truth, and the earlier lines have already shown that what *is* is a single undifferentiated whole, by this time the reader is in a position to understand that the subject is *all* that there is – *to pan*, the universe. It still remains true that Parmenides proposed no cosmology of his own in the Way of Truth, but his argument, I believe, aims to put constraints on any proposals future cosmologists may offer.

The second point of difficulty is the precise meaning of the various properties attached to what *is* in the passage. It is certainly compared to a ball in having spherical shape; but is any property other than shape being ascribed to it? It is sometimes claimed that the word that I have translated 'equally balanced' (*isopales*) means nothing more than 'equal,' and therefore asserts no more than the geometrical truth that all radii of a sphere are equal.[12] But it is a word that indicates equality of force. It is used of well-matched armies by Herodotus (I.82) and Thucydides (IV.94). The *pal-* root is akin to *palaiein*, 'to wrestle': Pindar used the compound *dyspales*, 'hard to wrestle.' There is good reason to believe, therefore, that

[12] For example, Guthrie, *HGP*, vol. II, pp. 43–4. There are many who take the same view.

Parmenides meant to ascribe not merely spherical shape but also a centrifocal dynamic pattern to his universe. Its one-ness, and stillness, are guaranteed by its balance of powers.

There are several striking similarities between these lines of Parmenides and the passage of Plato's *Phaedo* in which Socrates announces a new and strange theory of the earth's shape and position – a theory of which he claims to have been persuaded by an unnamed 'someone.'

> Well, I have been persuaded (he said) first that if it is in the middle of the heavens, being round in shape, then it has no need of air to prevent it falling, nor of any other similar force, but the likeness of the heaven itself to itself everywhere suffices it, together with the equal balance of the earth itself. For something equally balanced, placed in the middle of something all alike, will have no tendency to move any more or less in any direction, but being situated alike it will remain with no tendency to move. (*Phaedo* 108e–109a)

We find here too the geometrical conception of a sphere as the limit of all lines of equal length radiating from the center. Parmenides says of his subject it must be 'neither greater nor smaller ... in one place or another': Plato stresses 'the likeness of the heaven itself to itself everywhere.' Parmenides twice uses a word of the same root as Plato's 'likeness': Plato has the noun ὁμοιότης (likeness), the adjective ὅμοιος, and the adverb ὁμοίως; Parmenides has a noun ὁμόν ('its like') and an adverb ὁμῶς ('uniformly').

Now, there is no doubt that the theory of Plato's *Phaedo* is a centrifocal theory: it is offered as an explanation of why the earth remains stationary at the center of the universe without support from air or anything else. The verbal similarities we have noted confirm that we are on the right track in looking for the beginning of centrifocality in Parmenides. But there remains a paradox in finding any cosmological thesis at all in Parmenides' Way of Truth: his goddess explains that cosmology is all part of 'the deceitful *kosmos* of my words' (fr. 8.52). The Way of Truth cannot be about the position of the earth, since the earth is part of the illusory world of Seeming. But I do not think this is a compelling objection to what I am claiming. It has often been observed that the Way of Seeming is not totally unrelated to the Way of Truth. There is a hard core of concepts that is established in the Way of Truth and projected, so to speak, on to the more complex metaphysics of the Way of Seeming. Thus fire and night, the primary opposites of the Way of Seeming, each share some of the qualities ascribed in the Way of Truth to what *is*: the quality of self-sameness, for example (fr. 8.57–9). My notion is that Parmenides in the Way of Truth laid down a metaphysical framework into which he thought any cosmology would have to fit. One feature of it was the centrifocal balance of forces that he deduced from the perfection of what *is*.

The attribution of a centrifocal scheme to Parmenides gains confirmation from some features of the tradition already mentioned on p. 41. He was the first to declare that the earth is spherical: there is some confusion in the sources, but this statement rests on the reliable testimony of Theophrastus (cited by Diogenes Laertius VIII.48).[13] He was the first to discover that the evening star and the morning star are the same, said one source (Favorinus in Diogenes Laertius IX.21): again other claimants are mentioned, but Parmenides seems most likely. He knew that the moon's light is reflected from the sun, and may have been the first Greek to make this assertion in writing (Aetius II.8.5; and Parmenides fr. 14). As we saw in chapter 3, the sphericity of the earth is an almost necessary adjunct of the centrifocal scheme, and can hardly coexist with the linear picture; and these astronomical doctrines confirm that Parmenides had the right astronomy for a centrifocal system.

If I am right in seeing the beginning of the centrifocal theory in these lines of Parmenides, we have a striking case of a thesis about the physical world emerging from metaphysical speculations. As we shall see, the concept of atoms is another example. Later, the centrifocal universe was buttressed by arguments drawn from observation: the sphericity of the earth, for example, was inferred from the change in the stars visible in the sky as one moves north or south. A centrifocal dynamic system was then necessitated by the observation that the fall of heavy bodies is always vertical to the earth's surface and does not change its angle as one moves north or south. But initially, it seems, the inspiration of centrifocality was what one might call a metaphysician's intuition that what *is*, if it is to be bounded and complete, must be in perfect balance, and this can be achieved only if its quality and nature are exactly the same in all directions from its center.

It would seem to follow that we should also regard Parmenides as the first to demolish the idea of an infinite universe. His language seems to be definite enough: 'Strong Necessity / holds it in the bonds of Limit which girds it round about' (8.30–1); 'since there is an outermost limit, it is perfected / from all sides, like the mass of a well-rounded ball' (8.42–3). Yet there is something strangely inconclusive in his argument. The crucial point seems to be that what *is* must be *all* that there is: what *is* must not be deficient, or fall short. In temporal terms, this means that there is no time that is outside the duration of what *is*: what *is* has no beginning or end in time. In spatial terms, the same argument produces the conclusion that there is nothing situated outside the spatial continuum filled by what *is*.

[13] Burkert, *Lore and Science*, pp. 303–8, has a good discussion.

Parmenides' follower Melissus chose to interpret this requirement as meaning the same for space as for time: in magnitude, as in time, it is 'limitless' (*apeiron*, fr. 3), and Melissus' ground for saying this is just that it has no beginning and no end (fr. 2). From the same premiss, Parmenides argues that it is 'motionless in the Limits of great bonds' (fr. 8.26–8), and 'strong Necessity / holds it in the bonds of Limit' (8.30–1).

Both sides, in the controversy between the Closed World and the Infinite Universe, could shape this Eleatic argument to their purposes. We have seen the similarity between Parmenides and a passage of Plato's *Phaedo*, which takes up the finitist position. We shall see later how Epicurus echoes Melissus in his argument for infinity.[14]

5.3 *Pythagorean fantasies*

Another form of the centrifocal universe is described in our ancient texts. It belongs to the fifth century, but is in all probability later than Parmenides. This is the system ascribed by Aristotle to 'the Pythagoreans,' without naming individuals, but associated by other writers with Philolaus of Croton or Tarentum, who was a member of the Pythagorean school in the late fifth century.[15]

> Philolaus says there is fire in the middle around the central point, which he calls 'the Hearth of the Universe,' 'the House of Zeus,' 'the Mother of the Gods,' 'the Altar,' 'the Meeting House,' and 'the Measure of Nature.' Again, he says the periphery, highest in the universe, is another fire. But first by nature is the middle, and around this the divine bodies dance in chorus: the heaven <of the fixed stars>, the five planets, then the sun, under that the moon, under that the earth, under that the counter-earth, and under all of these the fire that occupies the position of the Hearth around the centre. (Aetius II.7.7)

> Philolaus the Pythagorean says fire is the centre, this being the Hearth of the Universe, second comes the counter-earth, and third the inhabited earth; this occupies and moves around in a position opposite to the counter-earth, for which reason the inhabitants of the latter are not seen by those of the former. (Aetius III.11.3)

This system is unique in antiquity in claiming that there is a center of the universe, but that it is occupied neither by the earth nor by the sun. Its motivation appears to come in part from a desire for numerological neatness.

[14] For further discussion of the Eleatic position with regard to infinity, see especially Cherniss, *Aristotle's Criticism of Presocratic Philosophy*, pp. 65–71; Fränkel, 'Studies in Parmenides,' in Furley and Allen, *Presocratic Philosophy*, pp. 25–36; Owen, *CQ* 10 (1960), 61–8.

[15] For discussion of the astronomy of Philolaus and other Pythagoreans, see Guthrie, *HGP*, vol. 1, pp. 282–301; Burkert, *Lore and Science*, ch. 4.

It is worth noting that Pythagoras himself, according to Aetius (II.1.1) was the first to apply the word 'cosmos,' with its implication of neatness and good order, to the world. Aristotle said the Pythagoreans were so entranced by parallels between number and nature that they invented the 'counter-earth' to make the heavenly bodies up to the perfect number of ten (*Metaphysics* I.5, 986a3ff.). He was probably right. Although the counter-earth featured also in explanations of eclipses of the moon (Aristotle, *De caelo* II.13, 293b15; Aetius II.29.4), the system as a whole makes very little astronomical sense, and it is hard to believe it was intended to do so. The face of the earth on which we live was said to be always turned away from the central fire and the counter-earth as it moved around the center, and yet the earth's motion was also said to be the cause of day and night. There is no explanation of what this means for the motion of the sun. The whole scheme lapses into fantasy when we are told that the moon is inhabited by non-excreting animals who are fifteen times more powerful than those on earth (Aetius II.30.1).

If this last idea seems absurd, another feature of the spherical Pythagorean cosmos of the fifth century was powerful enough to capture the imagination of poets and philosophers for many centuries, from Plato onwards. The stars made music in their orbits around the center. Aristotle gives a sober account:

> Some believe that sound must occur when bodies of such magnitude are in motion, since it is so with the movement even of earthly bodies neither equally bulky nor so rapid in motion. Sun and moon, and also stars so great in number and size, moving at such speed, must necessarily produce a noise of unimaginable volume. On these assumptions, and supposing that their speeds, determined by their relative distances, have the ratios of the musical concords, they say that the sound of the stars moving in circular orbits is harmonious. Faced with the apparent inconsistency that we do not hear this sound, they say that the explanation is that the sound is with us from the moment of our birth, and so it is not revealed by any contrasting silence, awareness of sound and silence coming from comparison one with the other. As sound and silence seem just the same to coppersmiths because they are so habituated, so it is with all mankind. (Aristotle, *De caelo* II.9, 290b21–9)

This idea appealed so much to Plato that he included it in the Myth of Er that concludes the last book of the *Republic*. Transposed into the Aristotelian mode, in which it is not the stars themselves that move but the spheres to which they are immovably attached, it passes into the western tradition as the Harmony of the Spheres.

The details need not concern us: it is hard to work out any set of speeds and intervals that might produce 'harmonious' music, let alone make sense as an astronomical theory. There is nevertheless much that is historically important in Pythagorean cosmology. It seems to be truly centrifocal: the orbiting earth could hardly be fitted into the linear directionality of the Milesian picture. It certainly applies mathematics to the physical world, and it attempts to find mathematical similarities between the physical structure of the cosmos and other things. Perhaps most significant is its determination to find *simplicity* in the mathematics of the cosmos. Relative speeds and distances, like musical intervals, are expressed as ratios of small integers, and the complex motions of the sun, moon, and stars are reduced to simple circles. The importance of the circle as a figure is brought out by the extraordinary religious value attached to the center – the House of Zeus, the Mother of the Gods, the Altar. Something of the same immensely powerful feeling was expressed by Kepler when he described his difficulty in rejecting the circle and substituting the ellipse in his astronomy. The circle remained the dominant figure for cosmology through all the centuries that separated the Pythagoreans from Kepler, for a variety of reasons. The scientific reason was that it proved possible to work out a model for the motions of the heavenly bodies, consistent with the available observations, based wholly on circular orbits. There was also an aesthetic reason, in the perception that the circle is a particularly beautiful figure in its 'all-round perfection,' noted by Parmenides. Finally, there were religious reasons, in that the perfection of the circle, and the fact that motion on a circular path can continue everlastingly without change, made it the most appropriate figure for the structure of the heavens, so long as the heavens were thought of as being either divine in themselves or the primary creation of God. The fifth-century Pythagoreans have a claim to be regarded as the first in the Greek tradition to fuse religious feeling with mathematics in their astronomy. The point is not simply that they both applied mathematics to their cosmos, and regarded the cosmic order as being divinely maintained: earlier Greek thinkers could claim so much. It is rather that they postulated a particular kind of cosmic order based on circles and concordant musical intervals from motives that were essentially religious and aesthetic.

As I have mentioned, there is a tradition that Pythagoras was the first to apply the word *kosmos* to the world (Aetius II.1.1), and that he did so because of the *order* observable in the world, especially in the motions of the heavens. Whether this is true or not, the word attests to the aesthetic value attached to the world order. It draws attention to the beauty of our world. The word *kosmos* was used from the earliest times to mean 'adornment' – for example, the elaborate trappings of a horse dressed for a fine occasion (hence our 'cosmetics'). But it refers particularly to the *orderliness* that is perceived as beautiful. The Pythagorean way of thought gave special emphasis to this aspect of cosmology. The world they thought of was finite, bounded by circles like the circular wall of a city. Inside the boundary there was harmony and good order. No doubt there will always be controversy about how much can safely be attributed to Pythagoras himself and his followers at various dates. But we can be sure that this school first sounded some of the themes that Plato later orchestrated in his cosmology, in which the world is presented as a single living organism, all parts contributing to a well-ordered whole and arranged on mathematically determined principles in the shape of a sphere. It is not difficult to understand why Plato, presenting his cosmology in dialogue form, chose as his spokesman the Pythagorean Timaeus.

6 Anaxagoras

In response to Parmenides, Anaxagoras worked out the astonishing theory that all physical change is nothing more than an emerging into view of something previously latent. Thus, we have no need to speak about what *is not*, but only about what we do not yet see.

The interpretation and assessment of Anaxagoras put forward here differ substantially from what is found in most contemporary histories of philosophy, in at least three respects. I believe Anaxagoras' theory is quite independent of Zeno, and that Zeno, in fact, criticized him. I also think that Anaxagoras did not criticize and modify Empedocles' theory of elements, but represents a more primitive stage of response to Parmenides than Empedocles. Moreover, my interpretation of his theory of matter is not quite the same as any other that I have read. These are all controversial positions, which cannot be defended in detail here.[1]

6.1 *Anaxagoras*

Anaxagoras began his book with a striking affirmation that at once made clear his opposition to Parmenides:

> All things were together, infinite in quantity and in smallness – for the small too was infinite. And of all things being together nothing was evident, because of smallness. (fr. 1)

Parmenides had agreed that what *is* is invariant in time, so that only 'it is,' and not 'it was' or 'it will be,' is correct, and that it is uninterrupted in space by any interval or change of density. He summed up this position in the words 'It *is*, all together.' Anaxagoras uses almost the same words, but marks the two essential differences from Parmenides by substituting plural for singular, and past for present. His ontology, as opposed to Parmenides', will allow one state of affairs to be distinguished from another

[1] A defense of this position can be found in 'Anaxagoras in response to Parmenides,' in my *Cosmic Problems*. The dependence of Anaxagoras on Zeno is argued in a special study by Raven, CQ 4 (1954), and it is a prominent theme in Calogero's book *Storia della logica antica*. The relation between Anaxagoras and Empedocles was discussed by O'Brien in *JHS* 88 (1968); he concludes that Empedocles wrote later than Anaxagoras. Schofield (*An Essay on Anaxagoras*, pp. 81–2) argues against my view.

in time – in other words, will allow *change* – and one thing to be distinguished from another in space. But this is to be done without admitting that any new thing ever comes into existence. Everything that now exists, or has ever existed, or will ever exist, has existed from the beginning of the world, when '*all things* were together.'

Change is rearrangement: as we have seen in chapter 5, this is the response of Anaxagoras to Parmenides' ban on coming-to-be. At the beginning of the world 'nothing was evident, because of smallness' – that is, the smallness of the *parts*. The model is the very simple one of mixing different substances together, like flour, milk, and eggs in a batter. One starts with three distinct substances, passes through intermediate stages in which portions of unmixed flour or milk or egg can be distinguished, and reaches a final stage in which these portions are broken up, and the mixture is homogeneous. The mixture still contains flour, milk, and egg, but now none of these is evident, 'because of smallness.' We must think of these steps in the process as taking place in the reverse order in cosmogony. The beginning of the world, Anaxagoras claims, was such a homogeneous mixture, containing all the ingredients that there are, and the growth of the world to its present and future conditions is a partial separating-out of the ingredients.

It is very important to notice how this scheme differs from Milesian cosmogony – at least, if we have taken the right view of the obscure Milesians in chapter 3. The Milesians postulated a primitive substance – water, the Boundless, air – that *changed into* or *came-to-be* differentiated substances. They saw, apparently, no problem in the *emergence* of earth, sea, and sky where there had been none of these before. Anaxagoras, too, held that there was a primitive, undifferentiated state of affairs from which the world grew; but led by Parmenides he insisted that the primitive state of affairs was a mixture of all the things that would later come out of it.

He differed from the Milesians in another highly significant way. This was the innovation that so excited Socrates, according to the passage of Plato quoted in chapter 2: '*Mind* set everything in order and is the explanation of everything' (*Phaedo* 97c). In fact, this idea was not original with Anaxagoras: he was anticipated by Xenophanes, who had spoken of a god – 'one god, greatest among gods and men, in no way similar to mortals either in body or mind' (fr. 23) – who 'without toil sets all things in motion by the thought of his mind' (fr. 25). But Anaxagoras made the idea a great deal more articulate. His cosmic Mind set all things in motion at a particular time and in a particular way. It may be that he was responding to the rhetorical question asked by Parmenides:

What need would have roused it
to grow, starting from nothing, at a later or earlier time? (fr. 8.9–10)

It is worth quoting the long, psalm-like fragment about Mind that luckily survives – one of the earliest surviving examples of Greek prose:

The other things have a portion (*moira*) of everything, but Mind is infinite (*apeiron*) and self-powered (*autokrates*) and is mixed with no other thing but is alone itself by itself. For if it were not by itself but were mixed with some other thing, it would have a portion of all things, if it were mixed with any; for in everything is a portion of everything, as I have said earlier. And the things mingled with it would hinder it, so that it had power over nothing in the way it does being alone by itself.

For it is the finest of all things and the purest, yes, and it has all knowledge about everything and has the greatest strength. Yes, and everything that has psyche, both the greater and the lesser – Mind has power over all of them.

And Mind had power over the whole circular motion, so that it moved in a circle in the first place. And first it began to move from a small beginning, and it moves more, and will move more in the future.

And the things being mixed together and separated off and distinguished – Mind knew them all.

And whatever things were to be in the future, and whatever things were that now are not, and whatever things are now and whatever things will be – Mind set them all in order; also this circular motion in which now move the stars and the sun and the moon and the air and the ether, which are separated off.

The circular motion itself made them separate off.

And there is separated off the dense from the rare, the hot from the cold, the bright from the dark, the dry from the wet.

There are many portions of many things.

But nothing is totally separated off nor distinguished, the one from the other, except Mind; but Mind is all alike, the greater and the lesser. Nothing else is alike, but each thing is and was, most evidently, those [things] of which there are most present in it. (fr. 12)

Many things in this charming and remarkable document deserve close attention, but we must confine ourselves to three topics, which will occupy the rest of this chapter. First, something must be said about the nature of Mind and its operations: what kind of cause does Mind represent? Secondly, we must examine in more detail the theory of matter and change. Finally, we shall say more about Anaxagoras' cosmogony, especially the traces of a mechanical theory of motion that can be found in it, and about his account of the development of life and civilization.

6.2 *Mind as a cosmic agent*

Socrates, in the speech quoted in chapter 2 (*Phaedo* 97c), castigated Anaxagoras for failing to make use of Mind so as to produce explanations mentioning why *it is best* for things to be as they are. Is that a justified complaint, or can we find a teleological theory in Anaxagoras?

It is true that in the fragments we find only a mechanical explanation of the position and motion of the stars, sun, and moon, and the other material components of the cosmos. 'I thought he would give an account of what is best for the particular thing, and of the general good,' said Socrates; but he heard no such thing from Anaxagoras. Clearly, Anaxagoras' Mind is a far cry from the Craftsman who in Plato's *Timaeus* builds the world with a mind always set upon what is best.

On the other hand, Anaxagoras did not choose Mind as his motive cause for nothing. If he had wished, he could have borrowed an agent from Parmenides' poem, and spoken of 'Necessity' (*Anankê*), a being which at that time carried no heavier burden of theory than the English 'cause' or 'force.' Instead, he chose a term that inevitably suggested intelligence, planning, and purpose. 'Mind set them all in order (*diakosmein*)'; the expression recalls the army commanders in the *Iliad*, who 'set in order (*kosmein*)' their troops, no doubt with an eye to what is best in battle. As we have already remarked, Anaxagoras' Mind recalls Xenophanes' god, who 'sets all things in motion by the thought of his mind,' and he in turn recalls Homer's Zeus, who 'sets Olympus in motion' with his nod. Anaxagoras himself treats Mind not only as a first cause of motion, but also as that which thinks and knows: it has 'all knowledge [or 'judgement,' *gnomê*] about everything,' it 'knew' (*egnô*; perhaps 'determined') all 'the things being mixed together and separated.' On the evidence of Plato, we must accept that there were no overtly teleological explanations to be found in Anaxagoras' book, but on the evidence of the language of the fragments we should refuse to put Anaxagoras firmly in the mechanists' camp. The idea of a purposive cosmos was there in embryo.

Development of the idea came soon afterwards. Diogenes of Apollonia in his book *On Nature* brought Mind down from its high eminence and grafted it onto a theory about air. Air according to him is the eternal substance that is the common origin of all things in the world, and air carries with it intelligence (*noêsis*). In differing forms it constitutes the souls of animals and man. More significantly, it is said to be responsible for tempering certain natural opposites: 'It could not have been so disposed without intelligence, so that there are measures of all things – of winter and summer, night and day, rain and winds and fair weather: and the rest, if one will remark it, will be found to be ordered with all possible excellence' (fr. 3). This seems to be the earliest explicit occurrence of the argument from design.[2]

[2] As to the dates of Anaxagoras and Diogenes, I believe that Anaxagoras' book was written shortly before 467, the date of the Aegospotami meteorite whose fall he was said to have

Air in Diogenes' theory is the material of which all the world is composed, and it carries intelligence with it everywhere, though in a different manner in every different kind of being. It is divine, it penetrates everywhere, and it 'steers' everything and 'disposes' everything. It plays a part, in fact, very little different from that of 'breath' (*pneuma*) in Stoic theory more than a century later, and we shall return to it in that context in volume 2.

6.3 *Anaxagoras' theory of matter*

A 'In everything there is a portion of everything' (fr. 12).

B 'Nothing is *totally* separated off nor distinguished, the one from the other, except Mind' (fr. 12).

C 'Each thing is and was, most evidently, those [things] of which there are most [parts? quantities?] present in it' (fr. 12).

D 'Of the small there is no smallest, but always a smaller ... also there is always a larger than the large' (fr. 3).

These four propositions give the essence of Anaxagoras' theory of matter. Although nothing in the later course of Greek philosophy is exactly like it, we can recognize in it the prototype of the continuum theory later adopted by Aristotelians and Stoics – and it is particularly interesting to notice how both the continuum theory and its rival the atomic theory (as we shall see), were brought into being in response to the arguments of the Eleatics.

We have already said that the point of Anaxagoras' theory was to provide an account of change that did not infringe the Eleatic ban on 'coming-to-be,' and the method chosen was to assert that all that emerged after the change was present before the change. At first sight, then, it appears that we should read proposition A in this sense:

A′ In everything [that changes], there is a portion of everything [it changes into].

A′ appears to be a statement of fairly limited range. Perhaps we could, in principle, list all the physical changes that are observed to occur and write down a limited set of statements of what contains what. We observe that

---

'predicted.' Diogenes is parodied in Aristophanes' *Clouds*, the original version of which was staged in 423.

Huffmeier, in *Philologus* 107 (1963), minimizes the teleological element in Diogenes, saying that he went no further than Anaxagoras. Fr. 3, however, seems to me to go significantly further.

beef, lettuce, wine . . . when eaten or drunk change into bone, blood, hair . . . and therefore we assert that the former contain portions of the latter. But we never observe rock, sand, iron . . . changing into bone, blood, hair . . . and so we do not, it appears, have to assert that the former contain any of the latter.

But this limited statement is not sufficient for Anaxagoras' theory. Taking each change in isolation is illegitimate, if the aim is to do away with coming-to-be altogether. We have to consider *all* the changes that may possibly occur in the infinite life of matter. Portions of rock may be washed away by rainwater, and from the silt there may grow plants, which may be eaten by animals, whose bodies return to earth . . . The rock, according to Anaxagoras' principle, must contain everything that may emerge from it at any stage in the natural cycle. In that case, it is difficult to think of any limitation that could be placed on the possibilities of change. Everything must contain *everything*. Anaxagoras intended to assert A, not the weaker A′.

If a piece of matter could be found that totally lacked some possible ingredient, then that ingredient could never emerge from it. Every such instance would put a stop to the possibility of change in the relevant respect. If anything existed in a condition of total isolation, then it could never change at all. Proposition B rules out this possibility, for all things except Mind. A substance may be separated from all the others to the extent that it becomes identifiable as that substance; but never in the history of the world is it totally separated from other things. Everything contains everything *always*.

But if everything contains everything, it follows that everything has exactly the same ingredients as everything else. In that case, how is anything at all to be identified? Proposition C answers this: although all the same ingredients are present everywhere, they are present, in the developed cosmos, in different proportions. Those things that preponderate are 'most evident,' and they give each thing its identity. Sweet red wine, although it contains smaller portions of everything, contains a preponderance of wine, the sweet, and the red (we shall comment later on the role of adjectival and other strange beings in Anaxagoras' ontology).

The complement of the principle of preponderance given in proposition C is what we might call the principle of *latency*. All the ingredients that are not preponderant and therefore 'evident' in a thing are to be thought of as latent in it.

This principle together with propositions A and B has the important consequence set out in proposition D: There is *no smallest*, and *no largest*. The first clause announces the principle of infinite divisibility, which is the essence of the continuum theory. Its place in the system is clear enough. If a thing *x* is latent in another thing *y*, it follows that *x* is smaller than *y*. Hence, if anything were 'the smallest,' it could have nothing latent in it; and the existence of such a thing is inconsistent with proposition A. The possibility of change in Anaxagoras' theory, as we have seen, depends on the principles of predominance and latency, and that is to say that it depends on there being always a relatively large and a relatively small. The existence of either a 'smallest' or a 'largest' would set limits to the possibilities of change.[3]

In order to get some intuition of Anaxagoras' theory of material change, one must eliminate the idea of atoms or molecules, or indeed particles of any kind. On p. 62 we used the example of flour, milk, and eggs, but a mixture of liquids provides the best model, so long as we forget our modern knowledge of their micro-structure. Imagine a mixture of different liquids, *a, b, c, d, e, f*... each having different perceptible qualities. The initial state of Anaxagoras' cosmos is a perfect and uniform mixture, such that every region has the same ingredients in the same proportions as every other. In this condition, all the liquids of which there is a relatively small quantity are undetectable, being 'drowned' by the more plentiful ones – in Anaxagoras' cosmogony, by air and aether. Then a rotation is started (more about that later), which has the effect of separating the constituent liquids somewhat, so that some of the liquids previously undetectable become predominant in one region or another. Perhaps *a* and *b* initially predominate everywhere: they are partially thrown towards the circumference by the rotation, so that *c* and *d* are left to predominate at the center. Every stir of the mixture may change the predominant character here and there. Such local swirls and eddies constitute the life of the cosmos.

The most difficult problem of historical scholarship on Anaxagoras is to determine precisely what are the constituents of the mixture. We have one list in a surviving fragment:

Before these things were separated off, all things being together [*sc.* in the original cosmic mixture], not even a color was evident. This was prevented by the mixture of all things – the wet and the dry, and the hot and the cold, and the bright and the dark, and much earth present in there, and seeds infinite in number, in no way like each other. (fr. 4)

It is obvious that this is not an exhaustive list, but a collection of samples. We have samples of the opposites, and one of the world masses (earth, water, air, fire), and in both cases it is a reasonable guess that the samples

[3] Thus there is no need to postulate Zeno's influence in explaining fr. 3.

are meant to represent the complete list. The word 'seeds,' here and in one other sentence in the fragment, has often been taken for a technical term in Anaxagoras, meaning 'particles' or 'portions'; but I think it likely that it has its normal meaning, and refers to the seeds of all the living species of the world. These too were in the original mixture.[4]

A tiresome problem, however, arises from the fact that Aristotle tells us plainly and repeatedly that in Anaxagoras' theory the original substances were 'the homoiomerous bodies.' This is an Aristotelian expression that gets its significance from Aristotle's own four-tiered analysis of the material contents of the universe. At the highest level of organization in that analysis we have (1) individual organisms, such as a man or a tree. These are broken down into (2) 'the non-homoiomerous parts,' such as face, hand, branch, leaf, which in turn are composed of (3) the 'homoiomerous parts,' such as bone, blood, skin, bark, wood. Finally, we reach (4) 'the primary bodies,' earth, water, air, and fire, of which the rest are composed. What characterizes the third tier and gives it its name 'homoiomerous' (meaning 'having similar parts') is that a part of one of these substances is synonymous with the whole. This contrasts them with (2): for example, skin has parts that are called 'skin,' but face is not made of parts called 'face' or 'faces.' It is worth noting, because it sometimes causes confusion, that it does *not* contrast them with (4): earth, water, air, and fire are homoiomerous, in Aristotle's sense, although he does not call them so.[5] Now what Aristotle says about Anaxagoras is intended to contrast him with Empedocles: unlike Empedocles, who made only substances of class (4) elementary, Anaxagoras held that class (3) are elementary. I believe that is all that he meant. Great confusion entered into the discussion because he was taken to mean that Anaxagoras held *only* the homoiomerous bodies to be elementary, or that he allowed something to be elementary if and only if it is homoiomerous in Aristotle's sense. But there is no need to give any of these stronger senses to Aristotle's words. What interested him in this connection was just that Anaxagoras did not derive substances like bone, blood, skin, etc. from anything more primitive, but held that they are all equally primitive, all being ingredients in the original mixture.

However, the list in Anaxagoras' fragment 4 does not contain *any*

[4] Contrast Guthrie, *HGP*, vol. II, p. 300: 'We may conclude that by a "seed" in fr. 4, Anaxagoras meant an imperceptible small bit of any kind of substance, that is, one containing that particular substance in greater proportion than all the others which of course are also in it.' I have explained my more limited view of Anaxagorean seeds in the article mentioned in note 1.

[5] See Aristotle, *De generatione et corruptione* I.1, 314a24ff.; *De caelo* III.3, 302a28; *Physics* III.4, 203a19. There is a sort of catalogue of Aristotelian homoiomerous bodies in *Meteorologica* IV.10, 388a13–20.

sample of the Aristotelian homoimerous bodies. It is a puzzling omission. But it can hardly be anything but an omission – it would be extravagant, just because of the silence of fragment 4, to reject Aristotle's testimony and the confirmation of many other ancient authors who had access to Anaxagoras' work. We must add to the list of members of the original mixture suggested by the fragment the homoiomerous natural tissues: hair, flesh, bone, blood, wood, bark, etc. etc.

Two points about this theory of matter deserve mention.

First, as we have repeatedly observed, it was designed as a means of explaining how change could occur in the physical world without introducing the irrationality of 'coming-to-be out of what is not'; and it succeeds in doing so. But that is all it does. It cannot be said to explain anything else at all, since it takes as its irreducible data *all* the ingredients of the physical world. Faced with the phenomena of nutrition and growth, it asserts that the tissues that grow were there all the time; to explain reproduction, it claims that the seed was in existence from the beginning, requiring only the addition of more of its component tissues to make it grow. Even primary perceptible qualities, such as colors, or hot and cold, are treated as substantive ingredients: the heat of a bath of hot water is not explained by the presence of fire in it, or by the motion or the shape or the properties of its particles, but is simply due to its containing a predominant amount of 'the hot.' Roughly speaking, the theory turns all the predicates that may be asserted of an object into ingredients of the object – but only roughly speaking, because it is easy to think of predicates that cannot be forced into this pattern. Predicates of shape, for example, and relative predicates, will not fit, and there is no clue as to how Anaxagoras handled them.

Secondly, it follows from the premisses of the theory, although it is nowhere brought into the open in the fragments, that we never encounter anything (except Mind) in its pure state. Everything in the world is a mixture, and the isolation of an ingredient is not merely something that has never been achieved: it is impossible on principle. This raises an epistemological problem: how can we know what gold is, if we can never encounter any better sample than a piece in which gold predominates over a vast, hidden junkyard of everything else in the world? The problem is not discussed in the fragments: it is merely asserted that Mind knows all the things that are in the mixture. I do not, however, bring this up as an objection to the theory: it is theoretically possible to infer from the evidence of the senses what a substance is and what properties it has without isolating it – indeed, physicists do it all the time. I bring it up in order to draw attention to a curious and very significant point. The fundamental, unchanging beings of Anaxagoras' world are directly accessible only to the Mind, not to the senses. The changing objects that we perceive around us have a portion of them, and if the portion is significant enough they take their names from them. The likeness of all this to Plato's theory of Forms is unmistakable, and we shall discuss it again in connection with Plato (below, pp. 171–3).

6.4 *The cosmos of Anaxagoras*

There is a well-known story in Plutarch's biography of Pericles that Anaxagoras, then a resident in Athens, was fined and exiled from the city, on a charge of impiety. The motive of the prosecutor was apparently to damage the Athenian political leader through his association with the atheistic foreign philosopher. The charge was specifically that he had called the sun a fiery stone, thus denying its divinity. In Plato's *Apology*, Socrates rebuts a similar charge against himself, saying that Anaxagoras had become the paradigm of an atheist and a materialist.[6]

We have already said something about the function of Mind in his cosmology. After the initial rotation imparted to a portion of the original mixture by Mind, we have Plato's word for it that Anaxagoras 'made no use of his Mind in forming explanations of the ordering of things, but invoked air, aether, water, all kinds of nonsense instead.' A page later than this, Plato broadens his complaint to cover 'the many' who put forward only matter-in-motion explanations: 'one surrounds the earth with a vortex and has it kept in place by the heavens, while another puts the air underneath it as a support, supposing it to be [shaped like] a flat breadboard,' instead of explaining how the earth is shaped and positioned *for the best.* Both the vortex and the supporting air are attributed to Anaxagoras by ancient authorities. Plato attacks a whole intellectual movement, but Anaxagoras is, for him, the best representative of the movement.[7]

Anaxagoras' cosmology is not fully described by the surviving ancient texts, but it is possible to get some idea of his position on the major controversies of natural philosophy.

There is no doubt that he thought the cosmos had a beginning in time: we have already translated fragment 12 (above, p. 63), which describes the initial movement imparted to the mixture by Mind – a small circular

[6] Plutarch, *Pericles* 32; Plato, *Apology* 26d.

[7] Plato, *Phaedo* 98b–c, 99b–c. Aristotle, *De caelo* II.13, 294b13, says Anaxagoras supported the earth on air. A rotation is guaranteed by the language of the fragments (fr. 12, fr. 13); this is distinguished from a vortex by Ferguson, *Phronesis* 16 (1971), but for obscure reasons. It may be that Plato has Empedocles in mind, rather than Anaxagoras, for this detail of the criticism, although Clement thought he meant Anaxagoras (DK 59A57). See below, note 10.

movement first, but gradually increasing in size. Does this movement eventually embrace the whole of the mixture, or are we to think of the cosmos as a limited part of the universe? Is our cosmos the only one of the kind, or are there others, now or at other times? Will the cosmos have an end in time, as well as having a beginning? The answer to the first question at least is reasonably clear: Anaxagoras says 'the surrounding' – i.e. the mixture that is around the cosmos on the outside – is boundless (fr. 2). But the other questions are more difficult.

There is a very interesting fragment that has often been taken to show that Anaxagoras believed in the existence of plural worlds. It is worth quoting at length:

> These things being so, it is right to think that there are, in all the things that are being put together, many things, of all kinds, and seeds of all things – [seeds] having forms and colors and savors of every kind. And [*sc.* these things being so, it is right to think] that men were compounded and other living creatures that have soul; and that by the men cities were settled, and farms established, *as is the case with us*, and that they have a sun and moon and the rest, *as is the case with us*, and that the earth grows much of every variety for them, of which they collect what is useful into their dwelling and make use of it. This, then, is my doctrine about the separation: that the separation would take place *not only with us, but anywhere else*. (fr. 4a)

The phrases that I have italicized may be thought to imply a replication of the cosmogonical process at other places and times in the universe. However, ancient writers, including Simplicius, who was probably able to read his book, class Anaxagoras with those who believed our cosmos to be unique. It seems best, therefore, to interpret the fragment as setting out a hypothesis – as indeed the opening phrase implies. *If* the initial mixture was as described, and things were gradually separated out by the rotation imparted by Mind, then 'it is right to think' that things were produced by the cosmogonic process just as we see them around us ('as is the case with us'). He is claiming that there is nothing incredible about the production of a cosmos such as we see ours to be, given the initial assumptions of his system.[8]

We may take it, then, that Anaxagoras belongs to the 'one world' camp – and since we have already dismissed the claim of Anaximander to have founded the other school (above, pp. 28–30), that means that the hypothesis of plural worlds still lies in the future. It also seems unlikely that Anaxagoras believed in a cycle of birth, death, and rebirth of the unique

[8] In this interpretation, I am following Fränkel, *Wege und Formen frühgriechischen Denkens*, pp. 284ff., and Vlastos, 'One World or Many in Anaxagoras?' in Furley and Allen, *Presocratic Philosophy*.

cosmos; if he had written about the death of the cosmos, our sources would hardly have failed to draw attention to it.[9]

The mechanism by which the cosmos came into existence was the sorting process brought about by a rotary motion in the mixture. Fragment 12, which we have quoted above on p. 63, describes how Mind starts and maintains the rotation: 'And first it began to move from a small beginning, and it moves more, and will move more in the future.' The heavenly bodies display the circular motion that was the cause of their being separated off, and 'there is separated off the dense from the rare, the hot from the cold, the bright from the dark, the dry from the wet.'

I believe this is the earliest use of the vortex or 'whirl' model for the sorting of matter in the cosmogonic process. As we have noted above, in chapter 3, note 7, the vortex is sometimes attributed to the Milesians, but although there is some slight evidence for that, it is probably wrong. The Milesians' picture of the qualitative change of a uniform primitive substance is different in kind from Anaxagoras' theory of the separation of ingredients in a mixture, and it is only the latter for which the vortex is a suitable tool. Empedocles made use of it for the same purpose, and the Atomists followed. It became a standard part of the matter-in-motion type of explanation of the cosmos.[10]

Certain kinds of vortex must be supposed to have been familiar to the readers of the early philosophers. They would include natural examples, such as whirlwinds, tornados, or 'twisters,' and artificial examples such as liquids stirred in a rigid container. These cases show a number of relevant phenomena. First and foremost, there is the circular motion. Secondly, heavy and large objects tend to collect at the *bottom* and in the *center* of the vortex; light, or less dense, objects tend to rise up the central axis of the vortex and then get dispersed towards the outside. (The effect of a twister on a tent and its contents makes a memorable illustration.)

Two points are of particular importance. First, the center of such a vortex is its *linear* axis: thus, it does not provide a model for the spherical Aristotelian cosmos, in which motion is focused on a central *point*. For-

[9] Simplicius, *Physics* 1185.9ff. and 154.29ff. implies that the Anaxagorean cosmos is permanent, once formed. Aetius II.4.6 denies it, but without evidence of careful research.

[10] My account assumes that Anaxagoras wrote before Empedocles, in agreement with Alexander, *In metaphysica* 28.1–10, Asclepius, *In metaphysica* 25.25–7, Kahn, *Anaximander*, pp. 163–6, and O'Brien in *JHS* 88 (1968), against most other commentators. As mentioned in note 7 above, Ferguson distinguishes between the περιχώρησις of Anaxagoras and the δίνησις or δῖνος of others. But it is clear that we have a process of sorting that is due to circular movement, and I do not understand what Ferguson wants to make of his distinction. The vortex model is well discussed by Tigner, *Isis* 65 (1974), with special relation to Empedocles. I have learned much from Mr Tigner's unpublished papers also, and am very grateful to him for letting me see them.

getting this has caused much confusion among commentators, ancient and modern. Secondly, the phenomena of the vortex are to a great extent determined by gravity; therefore, it does not provide a model for deriving or explaining gravity. The drift of large, heavy objects towards the bottom center is in part *caused* by gravity, and it will not serve as a stand-in for gravity. These two points taken together mean that as a cosmological model the vortex points to a unidirectional picture of the universe, in which weight is an absolute datum.

We can form a rough picture, then, of Anaxagoras' cosmogony. First, a small region of the original mixture is rotated by Mind, and this gives the light elements, air and aether, a tendency to rise up the axis of the rotation, and then to disperse to the outside, still carried around by the rotation. At the bottom of the whirling portion, the heavier materials tend to collect at the center, where they are held stationary, and form the flat disk of the earth. We must, of course, always be careful to speak of *tendencies* to separate, since, as we have seen, it is essential to Anaxagoras that separation is never complete. It continues only so far as to produce a predominance of certain substances in certain regions. There seems to be no physical explanation of this fact: Anaxagoras simply asserts it. It is, of course, a metaphysical necessity for his theory, as we have seen.

So far, the vortex model suggests a cosmos shaped rather like a modern observatory, with a flat floor, and a tall dome over the top of it, the whole being contained in an environment of boundless quantities of the unsorted mixture. This picture must be revised, however, in three respects.

First, the earth does not rest on the bottom, but is supported by air. We have already met this idea in the Milesians, but without the vortex. As a feature of the vortex theory, it is repeated by Empedocles, and we have some more detailed evidence there; so we will examine it in chapter 7.

Secondly, the flat disk of the earth does not lie at right angles to the axis of the vortex, but obliquely. This is a way of fixing up the vortex theory to fit the fact that the celestial pole is manifestly not vertically overhead. 'The stars at first moved as round a dome, so that the ever-visible pole was vertically above the earth, but later took a tilt' (Diogenes Laertius II.9). 'Diogenes and Anaxagoras said that after the cosmos was formed and had produced animals from the earth, it somehow tilted of its own accord towards the south' (Aetius II.8.1). No physical cause of this is given, and no reason for the odd idea that the tilt happened *after* the formation of animals. It is worth noting that, according to this report, it is not the earth that tilts, but 'the cosmos' – that is, the heavens; this is presumably because the overriding consideration was to maintain the direction of the fall of heavy bodies perpendicular to the earth's surface. If the earth tilted while the direction of fall remained the same, one would expect falling bodies to fall at an angle with the earth's surface.

The third complication is that the heavenly bodies themselves, whirling around the axis of the vortex at the circumference, were not regarded as light bodies like air and aether, but as stony objects uncharacteristically held up by the rotation. Since this doctrine, at the time, was held to be very scandalous and lacking in proper respect to divine beings, it is worth quoting from the ancient evidence about it. Plutarch describes it in his biography of Lysander:

> The stars are not in their natural place: stony and heavy, they shine by the resistance and tearing of the aether, and are dragged around forcibly held in position by the whirl (*dinê*) and pull of the circular motion, just as in the first place they were constrained not to fall to earth when the cold and heavy things were being separated from the universe. (*Lysander* 12 = DK 59A12)

It was the vortex that caught up these stony bodies at the beginning of the world and prevented them from joining the rest of the predominant earthy things at the bottom center, and it is the vortex that still whirls them around the sky, inflaming them by friction with the aether. This doctrine is associated with Anaxagoras' triumph in 'predicting' the fall of the meteorite at Aegospotami in 467, as well as with his disgrace and exile from Athens on the charge of impiety. In passing, we should note that it must count as the first explicit statement of a major plank in the anti-Aristotelian platform, namely, the claim that the heavenly bodies are of the same substance as the sublunary world.[11]

Anaxagoras' cosmos, then, is rather like a *tholos*-tomb hollowed out of a boundless mass of matter. Most of the cosmos is filled, without void intervals, with air and aether. The disk-shaped earth rests on a cushion of air on the floor of the *tholos*. At the circumference, all the matter is in circular motion, originally around a vertical axis but now around an axis that is tilted at an angle. Among the whirling matter are the heavenly bodies, stony objects that become heated to a glow by friction. The tilt of

[11] Some scholars suppose that Anaxagoras wrote after the fall of the meteorite, and used it as evidence for his theory. So Guthrie, *HGP*, vol. II, pp. 266 and 302–3; West, *Journal of the British Astronomical Association* (1960). But he was said to have *predicted* the fall. I think his theory of the heavenly bodies as stony was taken – perhaps by himself – to suggest the possibility that part of one of them might crash to earth one day.

Plutarch says that it is friction with aether that makes the heavenly bodies glow. Translated into Anaxagoras' theory of matter, I suppose this means that the aether somehow brings the fiery ingredients of the stony stars to the outside of their mass, so that fire predominates there and can be seen.

the axis means that some of the heavenly bodies, including the sun, moon, and planets, from time to time disappear from sight, when the whirl takes them behind the earth's disk.

Once the main world masses have been created, animal forms – presumably plants as well – began to grow 'from the moist and the hot and the earthy, later from each other' (Diogenes Laertius II.9). We have some testimony from Theophrastus: 'Anaxagoras says the air contains the seeds of all things, and they are washed down with the water and generate plants' (*History of Plants* III.1.4). There is a late report (Irenaeus II.14.2) that the same was true of animal seed. On this evidence, it is a reasonable guess that Anaxagoras used the knowledge that there are invisible airborne plant seeds to make plausible his theory that there are invisible seeds of all kinds in the mixture everywhere – seeds that grow, as plant seeds do, in a moist, warm, earthy environment. The theory that living forms first emerged from the earth, before the occurrence of sexual reproduction, is common to all the evolutionary cosmologists of Greece. They were left with the problem of explaining why the earth is not still observed to produce life spontaneously (or only to a limited extent). The Atomists had a solution to that, as we shall see, but there is no evidence that Anaxagoras tackled it, and no further information about the causes of change from spontaneous generation to parentage.[12]

There is one point of considerable interest here. As we set out the theory so far, we appear to have a sharp and unexplained transition from a *mechanical* process of sorting, caused by the Mind-initiated vortex, to a *biological* process of generation and growth. Since the explanation of such a transition still puzzles and even baffles contemporary scientists, it may seem striking that Anaxagoras passes over it in silence. From the language used by Anaxagoras, we can see, I think, that the transition did not appear so great to him. The word he uses for the 'separation' of the elementary substances of the mixture, *apokrisis*, is used in Greek biology to mean 'secretion'; it is applied, for instance, to the secretion of semen. Even the vortex can be connected with contemporary biological theory. It is used in an account of the development of the embryo in Hippocrates' *The Nature of the Child* 17; we shall return to this text in chapter 10. It seems clear enough, then, that Anaxagoras could not have thought of himself as

[12] That *some* creatures are regularly generated from the earth, not from each other, was a belief held by all Greek biologists. For Aristotle, see *De generatione animalium* I.1, 715a25, with Balme's note *ad loc.*, Balme's article, *Phronesis* 7 (1962), and Lennox, *Journal of the History of Philosophy* 20 (1982).

'reducing' life to chemistry and physics. There is not yet any sharp contrast between the behavior of animate and inanimate matter: or, to put it another way, we have not yet moved out of sight of the Milesians' hylozoism.[13]

This is important, and will help us later to understand the treatment of biological matter by the Atomists.

We have no connected information about the later stages of development in Anaxagoras' cosmological theory, and very little information of any kind. However, there exists a connected narrative of the stages of growth of the cosmos and of civilization, in the historian Diodorus Siculus. The precise sources used in this passage have been disputed for ages in the scholarly literature, and it would be a brave philologist who claimed it with certainty for Anaxagoras. But I think it is worth quoting here at length, to conclude this chapter, because if this is not what Anaxagoras wrote, it appears nevertheless to be fairly free of elements that can be identified as belonging to the later theory of Democritus and Epicurus, and it may be close to what he thought.[14]

At the time of the original composition of the universe, heaven and earth had a single form, their nature being mingled together. Afterwards, when their bodies parted from each other, the cosmos took on the whole arrangement now observed in it, the air got continual motion, and the fiery part of it hurried to assemble in the uppermost places – its nature being to move upwards because of its lightness. From this cause the sun and the remaining multitude of the heavenly bodies were caught up and enclosed in the whole vortex, whereas whatever was of the muddy and slimy sort, along with the assemblage of the moist natures, took their stand together because of their weight. And this, continually revolving in itself, engendered and created the sea out of the moist parts, and out of the more solid parts the earth, clay-like and altogether soft.

The latter at first solidified, as the flame of the sun shone upon it; then, when its surface fermented because of the heat, certain of the moist elements swelled up in

[13] A reminder that twentieth-century science is not entirely at ease on this subject: 'It is possible, of course, that life did not arise on earth at all. According to the theory of panspermia, which was popular in the 19th century, life could have been propagated from one solar system to another by the spores of micro-organisms. Francis H. C. Crick and Leslie E. Orgel recently made the more venturesome suggestion that the earth, and presumably other sterile planets, might have been deliberately seeded by intelligent beings living in solar systems whose stage of evolution was some billions of years ahead of our own' (Dickerson, *Scientific American* (September 1978), 70).

[14] Dodds, *The Ancient Concept of Progress and Other Essays on Greek Literature and Belief*, p. 11: 'In an earlier (unpublished) version of this paper I argued at some length that if we had to name a single source for the whole the likeliest was in fact Anaxagoras or his pupil Archelaus. But I should now be content to suggest that after all we should take Diodorus at his word and assume that he, or more likely some Hellenistic predecessor, being

many places, and round about them grew pustules covered by delicate membranes – a thing still observed occurring in marshes and swampy regions, whenever the ground becomes cold and then the air suddenly becomes very warm without undergoing a gradual change. The moist elements were impregnated with life by means of heat, in the manner described, and by night they at once took in nourishment from the mist that fell from the surrounding air, while by day they were made solid by the heat. Finally, when the embryos had attained full development and the membranes were heated through and burst open, there grew all types of animal life.

Of these, those that had the biggest share of heat departed to the upper regions having become winged, those that retained an earthy composition were counted in the ranks of creeping things and the other land animals, and those that had an especially large share of the moist nature gathered into their congenial region, and were called water creatures. But since the earth grew continually more solid because of the heat of the sun and the winds, in the end it was no longer able to produce any of the larger animals, but living creatures of every kind were generated by intercourse with each other.

Now concerning the genesis of the universe this is the account we have received. The first men to be born, they say, having a disorderly and beastly way of life, went forth one by one to get food and gathered the tenderest green plants and the wild tree fruits. Being attacked by wild beasts they came to each other's aid, being taught by expediency; assembling together through fear they gradually came to recognize each other's characteristics. Their voice being at first unintelligible and confused, they gradually articulated their utterances, and by establishing with each other symbols for everything they encountered, they set up a mode of expression concerning all things. But since such communities grew up all over the inhabited world, they did not all have the same language, since each group organized its words as chance would have it. That is why languages of all kinds exist. And the first communities became the ancestors of all the peoples.

So the first men had a hard time of it, since none of the things useful for life had been discovered. They had no clothing, they were unacquainted with housing and fire, they were totally ignorant of cultured food. Since they did not know how to harvest wild food, they made no stores of fruits to cater for their needs. Hence many of them perished in the winters through cold and lack of nourishment. Gradually, however, being taught by experience, they resorted to caves in the winter and they stored such fruits as could be preserved. When they had gradually learnt about fire and other useful things, the crafts were discovered and the rest of

---

no philosopher, consulted a doxographical manual and out of what he found there put together a not very up-to-date summary of the opinions most often attributed to rationalist thinkers.'

There is a close study of this text in Cole, *Democritus and the Sources of Greek Anthropology*.

The section on the origin of language seems to me most likely to be Hellenistic.

what could benefit their societies. For to speak generally, necessity itself was man's teacher in all things, giving appropriate instruction in everything to a creature well endowed by nature and equipped with hands, speech, and quickness of mind to assist in all tasks. (Diodorus 1.7–8)

7 Empedocles and the invention of elements

7.1 *The four roots*

It is customary to regard the Sicilian Empedocles as a particularly archaic and reactionary thinker, but I believe that is a mistake. Like Parmenides, he wrote in hexameter verse, and he chose the language of myth. He adopted the Pythagorean idea of metempsychosis, and wrote of himself as a fallen *daimon*, one who had been 'a bird, a bush, and a dumb sea-fish.' Historians of science find it hard to take someone like that seriously. Nevertheless, he was responsible for three innovations of great importance to the history of science: the invention of *elements*, the postulation of two physical forces, qualitatively different from each other, and the first statement, so far as we know, of the concept of natural selection by survival of the fittest.

His theory of elements claimed that there are just four substances in the physical world: earth, water, air, and fire. There is a finite and unvarying quantity of each of them, and between them they make up all the material objects that there are, by mixing in different proportions. Empedocles gives an idea of how the theory works by offering a comparison with four-color painting:

> As when painters are decorating votive tablets –
> men well skilled and cunning in their craft –
> they take the many-colored pigments in their hands,
> mixing in due proportion more of some and less of others,
> and from them construct forms in the likeness of all things,
> creating trees, and men, and women,
> beasts and birds and water-nurtured fishes,
> yes, and gods endowed with long life, highest in honors:
> so let not deceit persuade your mind that any *different* source
> has brought forth mortal things, countless, to the light.[1] (fr. 23)

[1] The participles describing the painters in line 2 are *dual* in number. Perhaps that is of no significance, as Bollack thinks (*Empédocle*, vol. III, p. 122). But possibly Empedocles is anticipating the cashing of his simile, and is thinking of Love and Strife.

It is known from other sources (Theophrastus, *De sensibus* 73 = DK 68[Democritus] A135) that Greek painters worked with four basic colors. See Guthrie, *HGP*, vol. I, p. 148, n. 1. Aetius says Empedocles noted the colors are equal in number to the elements – viz. white, black, red, yellow (I.15.3 = DK 31A92). The 'black' pigment must have been a very dark blue, if painters were ever to get blue.

'Any different source' may mean a mode of generation different in kind from the mixing of the four basic colors that produce painted forms, or it may perhaps refer to preceding lines, now lost, and mean a source other than the four elements. The general sense in either case is clear. Just four colors are known to produce all the varied objects that clever painters can put into their paintings: so we must believe that just four elements can produce all the variety of the physical world.

It is important to distinguish Empedocles' thesis from others that have some similarity. He differs from the Milesians in that they operated with a changing basic substance, whereas his four elements were unchanging. This is the difference brought about by Parmenides' criticism of the notion of change, and we have already mentioned it in section 4.4. We have also already mentioned there that he differs from Anaxagoras in limiting his original substances to four, and allowed that compound beings in a sense 'come-to-be,' whereas Anaxagoras tried to ensure that in nature nothing comes-to-be that has not always been. This is an important step in the direction of atomism, and we shall return to discuss it later (see chapter 9). He differs from Plato, Aristotle, and the Stoics, who also adopted the four elements into their physics, in that whereas his earth, water, air, and fire were permanent, enduring physical bodies, later physics treated them as changing into each other. Like the Atomists, Empedocles interpreted all change as rearrangement of enduring substances: Aristotle and the Stoics held that no physical substance in the sublunary world persists forever in the same state (Plato does not fit either pattern exactly).

One can feel confident, then, in ascribing to Empedocles the invention of elements, although of course he did not invent or discover earth, water, air, and fire. For the first time, he made the claim that physical objects can be divided into a class consisting of simple bodies and a class consisting of compounds made of simple bodies. He claimed too that each of the four simple bodies has its own characteristic properties, and that the properties of compounds can be wholly explained as derived from the properties and the arrangement of the simple bodies. He asserted explicitly (in fragments 13 and 14) that there is no void space in the universe, probably on the ground that to talk about void space is to talk about *nothing*, and that is to talk nonsense – an Eleatic argument to be found in the surviving fragments of Melissus (fr. 7.7; see below, chapter 8). In Empedocles' theory, therefore, all properties of compounds must be derived from properties of elements: they cannot be explained, as they were later in the atomic theory, by the proportion of matter to void space. Thus, earth gives solidity to a compound, water gives fluidity or pliancy, and so on. The properties of the simple bodies belong to them unalterably: if a pond freezes hard, we must interpret that not as a case of the element water losing its fluidity, but as an incursion of earthy matter into the pond.

The elements are thus endowed with characteristics corresponding to those perceived in the familiar items in the physical world that bear their names. But they are idealized forms of these familiar beings. They even have divine names:

> First, hear of the four roots of all things:
> shining Zeus, Hera the life-bringer, Aidoneus
> and Nestis who wets the mortal spring with tears. (fr. 6)

Like the gods of Homer, they live forever, they retain forever their characteristic powers, and they have these powers in a higher degree and purer form than their earthly counterparts. As well as being gods they are also called *roots*, in the very simple sense that things grow from them.

Why did Empedocles choose just these four as the elements? Up to a point, they represent different states of matter: solid, liquid, gas – but fire does not fit clearly into this sequence. The four can be fitted into a pattern made by four primary qualities, hot and cold, dry and wet. Thus fire is hot and dry, earth cold and dry, air warm and wet (steam was counted as a form of air), and water cold and wet. This is a neat pattern, used later by Aristotle, but there is no evidence that Empedocles had thought of it. It does not appear intuitively obvious, since one has to do some thinking in order to fit air into the pattern. The clearest characteristic of the four is perhaps their natural position in the world. Earth lies at the bottom, and water, in the form of seas, lakes, and rivers, lies on the top of it. Air is above the earth and water, and fire, represented primarily by the sun, is above the air. There is no obvious possibility of reducing the number four by identifying any of these with any of the others, and no obvious need for any fifth body to be added to the list; Aristotle's fifth body was postulated to satisfy the requirements of a sophisticated theory of motion. Without attributing to Empedocles anything so precise as Aristotle's doctrine of natural place, we may guess that among the considerations that led him to choose his four elements was their disposition in the world order.

Before going on to Empedocles' account of the formation of our world and its contents (a highly controversial subject), it may be useful to review briefly what evidence there is for his derivation of compounds and their qualities from the elements.

But come, behold this witness of our earlier words
if anything in those earlier words was insubstantial in form;
behold the sun, bright to see and hot everywhere;
[?] is bathed in shining radiance;
behold the rain, in all things dark and glacial;
and from earth comes issue that is close-packed and solid.
Under Strife, they are distinct and separate,
but in Love they come together and are desired by each other.
For out of these come all things that were and are and will be;
trees grow, and men, and women,
beasts and birds and water-nurtured fishes,
yes, and gods, endowed with long life and highest in honors.
There *are* just these very things: but passing through each other,
they become different in appearance – so much does mixture change them.
(fr. 21)

This fragment, like many others, stresses the permanence of the elements and their ability to produce from themselves the whole range of perceptible objects in the world. There is a brief mention in the opening lines of the qualities associated with the elements: fire (the sun) is *bright* and *hot*, water is *dark* and *cold*, earth is the origin of *solidity* (line 4 is textually corrupt and very obscure: it should refer somehow to air). Either Empedocles himself or the vagaries of citation by surviving authors have left us with some other curiously assorted scraps of information: shellfish and tortoises (called 'stoneskins') owe their characteristics to a quantity of *earth* in their upper parts (fr. 76); fish, being watery, are more fertile than creatures of the land or air (Plutarch on fr. 74); plants (evergreens?) get certain properties from the air. Aristotle, in *Meteorologica* IV, points out that all woody material, when charred, gives off smoke, and he quotes Empedocles as classifying hair, leaves, feathers, and scales as being 'all the same' – presumably all earthy. On the structure of bones, we have these lines, surely among the strangest documents in the history of biochemistry:

Kindly Earth, in its broad-chested melting pots,
took two parts of gleaming Nestis [*sc.* Water] out of eight,
and four of Hephaistos [*sc.* Fire]: these became white bones,
divinely fitted together by the adhesives of Harmonia [*sc.* Love]. (fr. 96)

And this, on the composition of blood:

Earth encounters these, virtually equal to them –
encounters Hephaistos, rain, and all-shining Aither –
coming to anchor in the perfect havens of Love,
whether a little greater or much less than them;
from them comes blood, and other forms of flesh. (fr. 98)

The last two quotations make clear something that could well have been inferred from the principles of Empedocles' theory, that the proportion of the ingredients is important. The reasoning behind the choice of proportion is partly intuitive common sense (obviously hard solids have more earth in them, and so on) and partly *a priori*: the one-to-one formula for blood, for example, is probably connected, somehow, with the fact that he took blood to be the organ of thought – perhaps thought requires that its organ be unprejudiced or balanced.

7.2 *Love and Strife*

I will tell a double theme. Now, it grows to be one alone,
out of many; now it grows to be many again, out of one.
Two-faced is the birth of mortal things, two-faced their passing:
the one is born – and destroyed – by the coming together of all things:
the other, fostered by their growth again apart, is also thus dispersed.
These things never cease from continual exchange,
at one time coming together through Love, all into one,
then again carried each apart, through the hatred of Strife.
Thus, in that *one* has learned to grow out of *many*,
and again, when *one* grows apart, *many* come into being,
in this way they come-to-be, and their life is not for ever;
but in that they never cease their continual exchange,
in that way they *are* for ever, unchanging in their cycle.

But come, listen to my words: learning will expand your mind.
For as I said before, declaring the bounds of my words,
I will tell a double theme. Now, it grows to be one alone,
out of many; now, it grows to be many again, out of one:
Fire and Water and Earth and the vast height of Air –
and Strife the destroyer, at odds with them, equal everywhere,
and Love, implanted in them, equal in length and breadth.
Look at *her* with your mind: sit not with dazed eyes.
She is respected by mortals, implanted in their limbs,
through her they think loving thoughts and do kindly deeds,
calling her Joy and Aphrodite by name.
No one has seen her with his eyes as she goes about,
no mortal man: but *you* must hear my undeceiving course of words,
for these [*sc.* the elements] are all equal and of equal birth,
but they have different honors, and each his own character,
and they rule by turns, as time circles around.
And to these nothing is added, nothing passes away from them;
for if they perished utterly, they would no longer be;
and what could give increase to the ALL? And from whence?
How could they perish, since nothing is empty of them?

> No, there are just these very things; but running through each other
> they become different things in time – and are ever and always the same.
> (fr. 17.1–35)

This fragment may be the first introduction of Love and Strife as characters in the poem, and since it sets out many of the principles of Empedocles' theory – and exhibits many of the problems – it seems worth quoting at length.

The main theme – 'a double theme' – is an assertion of his response to Parmenides, and an explanation of fragment 9 of his own poem:

> When they are mingled together to form a man, then men speak of coming-to-be,
> and when they are sundered, then of ill-fated death.
> They are right to call them so, and I follow the custom in so speaking.

When the elements, which are ungenerated, indestructible, and unchanging, form a compound, then that compound – a unity – can be said to come-to-be; when the compound splits up into its components again – a 'many' – the compound can be said to die or pass away, and particular quantities of the elements can be said to come-to-be. The generation of the One is the destruction of the Many, and vice versa – but 'generation' and 'destruction' in this sentence are convenient and customary locutions, to be interpreted by the philosopher in terms of mingling and separation.

The exact subject-matter of fragment 17 is disputed. Some interpreters believe that the One that is said to grow out of many is the cosmos, or includes the cosmos, others that it is only a specimen mortal compound. For the present we can keep an open mind about that problem, and treat the fragment as being about the birth and death of *any* compound, without prejudice to the question of whether that includes the cosmos or not. We shall return to that question in section 7.5.

What exactly is the ontological status of Love and Strife? Empedocles uses some very curious language about them: Strife is 'equal everywhere,' Love is 'equal in length and breadth.' Love can be called by the personal name, Aphrodite. It is almost as if there were six elements, not four:

> By earth we see earth; by water, water;
> by aether, bright aether; by fire, destructive fire;
> and love by love, and strife by dreadful strife. (fr. 109)

So some have even wondered whether the individual psyche, which Empedocles believed to survive the death of the body, might be composed, in his theory, of a 'parcel' of Love and Strife, or, in especially meritorious cases, of Love alone. Yet it seems doubtful that it can be right to put Love and Strife on such an equal footing with the four physical elements. When Empedocles wrote that Love is 'equal in length and breadth,' he probably meant that it is coextensive with the elements, in the sense that there is no part of the world that does not feel its power. His language is still close to the language of the Homeric epics, in which words have wings and oaths are broad. The language for distinguishing physical objects from their properties and relations had not yet been created: we may recall Anaxagoras' use of 'the hot,' 'the cold,' 'the large,' 'the small,' etc. But because his language was blunt-edged, we need not conclude that he made *no* distinctions. Like everyone else, Empedocles no doubt realized clearly, in spite of the implications of his language, that a couple in love is a couple, not a trio with Love as a third member.

If we try to arrive at a precise determination of the roles of Love and Strife in Empedocles' cosmology, we run into some difficulties. There is enough evidence, as we have seen in fragment 17, to prove Love's role in the uniting of different elements into a compound, and Strife's role as the destroyer of the compound. But several questions remain to be answered. Is unification by Love the only method of formation of an organism, or is there a reverse generation in which organisms are formed when homogeneous wholes are (partially) disrupted by Strife? My answer to this is that unification by Love is the only cause of the formation of organisms; but the reasoning for this answer must be postponed to later sections. The second question is this. Is the work of Love confined to the formation of compounds, or is Love also responsible for the like-to-like attraction between parts of the same element? To give a particular (and crucial) example: is the location of the masses of earth, sea, and the heavens the work of Love, or of Strife? This turns out to be a very difficult question. Without much confidence, I am inclined to believe that Empedocles meant Love to be only a unifier of different elements. The metaphor is plainly a sexual one: its origin lies in the Homeric formula for love-making, 'to mingle in love.' Self-love is sterile; only heterosexual love produces children. On the cosmic scale, there is generation only where the elements are mixed with each other; even the earth generates life only in the presence of moisture, air, and heat. If Love has a role in the shaping of the cosmos as a whole, as opposed to the making of its organic parts, I suggest that it must be in keeping it together so that the elements can mingle where they are contiguous. Without Love, the elements would separate completely. Thus the absence of void in the cosmos, if this is right, is not a constraint on the action of Love and Strife but a consequence of the force of Love.[2]

The figures of Love and Strife stand in an ambiguous position, poised between myth and science. Their names, and the language used by Empedocles to describe them, come from the realm of myth. As we have seen, the activity of Love in her cosmic role can be seen as an elaboration of the Homeric metaphor of 'mingling in love.' They are easily personified, and indeed they are very recognizable in the persons of Venus and Mars in the famous prelude to Lucretius' *De rerum natura*. And yet they can also be perceived as the ancestors of the forces of attraction and repulsion in modern physics. The extent to which this is true differs according to one's view of the exact nature of their roles in Empedocles' physics. As I have said, some think that Love is responsible for all attraction, even between like things, and Strife for all repulsion; others, like myself, regard Love as that which attracts and unifies unlike things, and Strife as the force that marshals like things together like a national army in opposition to a foreign foe. In either case, the struggle between Love and Strife is a far cry from the succession myths of the old Theogonies. The action of Love and Strife is not arbitrary and imperious, but regular, predictable, and even, to a certain extent, observable. They have gods' names, but they are not supernatural. Even more, perhaps, than Anaxagoras' cosmic Mind, they are the forces of nature itself, identified by Empedocles by a brilliant stroke of generalizing abstraction from the observation of nature.

7.3 *The formation of the cosmos and its contents*

We have now seen how Empedocles aimed to explain the generation and destruction of individual items in the cosmos, and the perceptible qualitative changes that take place around us. As Aristotle would later say, he distinguished the material causes of things, namely earth, water, air, and fire, with their associated qualities, and the efficient causes, Love and Strife. About these aspects of his philosophy there is a wide measure of agreement among scholars. However, in addition to this account of the generation and composition of individual things, Empedocles joined with the Milesians and Anaxagoras in offering a theory about the origin of the cosmos as a whole and its development by stages into the state in which we observe it now. And this is much more difficult and controversial.

Some points are clear. Empedocles spoke of a condition of the universe in which everything that exists is gathered together, through the influence

[2] See section 7.6 'Appendix: The action of Love and Strife.'

of Love, into a single whole in which no difference is discernible – a striking parallel with the initial condition of Anaxagoras' world, 'all things together.' He called this being a god, and gave it the name 'Sphairos' – a masculine version of the ordinary Greek word for a sphere. This was not the first instance of a single, all-embracing cosmic deity in the history of Greek thought: Xenophanes of Colophon, who emigrated to Sicily and might even have met Empedocles there in his old age, had spoken of 'one god,' who 'sees as a whole, thinks as a whole, hears as a whole' (fr. 24), and 'remains always in the same place, without moving' (fr. 26) while organizing the motion of the cosmos. Like Xenophanes, Empedocles had to explain that this cosmic deity was not of the same order as the Homeric gods:

> From the back, no twin branches spring,
> there are no feet, no swift knees, no generative parts:
> but it was Sphairos, everywhere equal to himself. (fr. 29)

> But he is equal everywhere, and boundless as a whole,
> Sphairos, rounded, rejoicing in circular stillness.[3] (fr. 28)

The ancient authors who quote these lines make it clear that they describe not a separate, supramundane deity like the One God of Xenophanes, but a particular state or aspect of the universe; and one set of Empedocles' verses quoted by Simplicius confirms this, by mentioning the sun:

> then neither is the swift body of the sun discerned.
> .....
> Thus everything is held tightly in the close darkness of Harmony,
> Sphairos, rounded, rejoicing in circular stillness. (fr. 27)

In the universe in this condition, Strife plays no part; but at some time – no reason is given why it happens at one time rather than another – the perfect harmony of Sphairos is disrupted. 'As soon as Strife begins to prevail,' says Simplicius, 'motion takes place again in the Sphere:

> For all the limbs of the god were shaken one after another.'[4] (fr. 31)

This is a stage of the proceedings of which no good description survives.

[3] The word translated 'stillness,' μονίη, has sometimes been taken to mean 'solitude' or 'oneness.' There is a good discussion, defending 'stillness,' in Guthrie, *HGP*, vol. II, p. 169, n. 3.

[4] The *Stromateis*, ap. Eusebius, *Praep. evang.* 1.8.10 = DK 31A30, says the revolution began (?) 'when fire used its weight in the direction of its greatest concentration (ἀπὸ τοῦ τετυχηκέναι κατὰ τὸν ἀθροισμὸν ἐμβρίσαντος τοῦ πυρός)' – I follow Bollack's interpretation of a difficult clause (*Empédocle*, vol. III, p. 219). This is a detailing of Strife's work, not a rival cause to Strife.

There are three sketchy accounts, in the *Stromateis*, Aetius, and Philo (DK 31A30, 49), which agree in outline: first the aether (that is, the brighter upper air) was separated out from the mixture and distributed 'in a circle'; secondly fire escaped, and 'having no other place to go ran out upwards, below the fixed mass (*pagos*) that is round the air' (*Stromateis*); thirdly, earth was formed, and 'when earth was tightly constricted by the rotation, the water gushed out' (Aetius), and the lower air was given off as vapor by the water.[5]

There is some evidence about the end of the process of separation. The best testimony is in Plutarch's little dialogue, *The Face on the Moon* – unfortunately nothing really informative has been preserved among the fragments of Empedocles' own work. One of Plutarch's characters wants to argue against the thesis that the motion of the four elements and their position in the cosmos are entirely due to their nature: if everything takes its natural place, he argues, the awful result may be the dissolution of the cosmos, as under the power of Strife in Empedocles' theory.

> Earth had no part in heat, water no part in air; there was not anything heavy above or anything light below; but the principles of all things were unmixed and unloving and solitary, not accepting combination or association one with another, but avoiding and shunning one another, and moving with their own peculiar and arbitrary motions, they were in the state in which, according to Plato [*sc. Timaeus* 53b], everything is from which God is absent, that is to say in which bodies are when mind or soul is lacking. So they were until desire came over nature providentially... (Plutarch, *De facie* 926E–F, trans. Cherniss, slightly adapted)

There are many implications in this passage. It is implied that the separate existence of the four elements is a state that is not instantaneous but has some duration. In this state they are thought of as moving, not stationary – but with self-willed movements (φορὰς ... αὐθάδεις). This seems to me to preclude the picture of the elements spinning in regular concentric spheres around an axis.[6] It is also implied that in keeping themselves to themselves and refusing to associate with each other, the elements were demonstrating a lack of love (they were *astorgoi*). This seems

[5] The *pagos* is the crystalline outer shell of the cosmos, according to Aetius II.11.2 (= DK 31A51). See O'Brien, *Empedocles' Cosmic Cycle*, pp. 287ff.

[6] O'Brien, *Cycle*, pp. 146–56, adopts the picture of concentric spheres spinning round an axis. His argument for the spherical shape, however, is extremely perfunctory. It amounts to no more than the observation that natural place, for Aristotle and for the speaker in Plutarch's dialogue, consists of four concentric spheres. So it does. But the point of citing Empedocles, both in Aristotle's *De caelo* 301a11–20, quoted here by O'Brien, and in Plutarch, is quite plainly not the *shape* but the *separation* of the elements and the fact that they were moving.

to me to tell against the claim that not only the association of different elements, but also the conglomeration of a mass of the same element, is the work of Love.[7] Another implication worth noting is that even in this stage of development of the universe, the elements are described as *heavy* or *light*. These are not properties that are acquired only in the cosmic stage. The heavy is all 'below' and the light 'above': but nothing is said or implied about whether 'below' means 'at the center' or 'at the bottom.'[8]

The activity of Strife, then, arranges the four elements in masses, separate from each other. This makes the rough structure of our cosmos. It is important to observe this: the final outcome of Strife's work is not destructive of cosmic order in the sense of putting everything into a chaotic mêlée. The four elements in the cosmos in which we live are, in fact, very much where Strife put them: a mass of earth at the center, with water, more or less massed in the oceans, on top of it, air and the fiery heavens around and above. In this sense it is true that according to Empedocles Strife created our cosmos, as some of our sources imply. But Plutarch is right too, of course, in that the *life* of our cosmos depends on mixture, and the separating work of Strife, if carried to the limit, makes life impossible.

I have been careful in this section to distinguish between 'the universe' and 'the cosmos.' It is sometimes thought that in Empedocles' system it is precisely the same quantity of material that makes up, at different stages of development, the Sphairos, the scheme of elements in isolation from each other under the power of Strife, and the cosmos in which we now live. The cosmos is after all, according to this view, nothing but the Sphairos rearranged internally by the struggle between Love and Strife. I want to maintain that Empedocles did not belong to the spherical school of cosmologists; and as a preliminary to that, I shall argue against the identification of the Sphairos with the cosmos.

There is just one piece of relevant testimony in favor of this suggestion. Aetius says: 'Empedocles holds there is one cosmos; the cosmos is not the universe, however, but is a small part of the universe, while the rest is inert matter' (Aetius 1.5.2 = DK 31A47). If this is believed, it is conclusive: no more is needed. The Sphairos is, of course, the universe – nothing but the universe could be said to be 'entirely boundless' (fr. 28). Within it, we must suppose, Strife stirs up trouble in a limited area and *in that area* separates the elements from each other in the manner we have just discussed; and *in that area* Love reunites the elements to make our cosmos, as we shall see. If this is correct, it puts Empedocles in the mainstream of

[7] See above, section 7.6.

[8] O'Brien has to assume that it means 'at the center,' since he is committed to the spherical model.

Presocratic tradition, from Anaximander on through Anaxagoras and (with a difference) to the Atomists. In a three-dimensional 'field,' which can be described as 'boundless,' the cosmos is a hollow shape, bounded by some kind of shell, with earth on the middle axis, more or less at the bottom of the hollow, the rest of the hollow above and below the earth being filled with air.

This report of Aetius is usually rejected, or explained as referring to a time before our cosmos was fully formed.[9] If it is rejected, it is generally because it is taken for granted that the 'inert matter' would have to be outside the Sphairos. I agree that there are good reasons for rejecting that. But is there any compelling reason to reject the notion that a part of the Sphairos is left outside the cosmos? Certainly the Sphairos itself is said to be destroyed or disrupted when the cosmos is formed, and in fragment 29 Empedocles uses a past tense in speaking of it; but that is perfectly compatible with *some* of the Sphairos being left undisturbed. It is disrupted because the formation of the differentiated and moving cosmos within it puts an end to its total homogeneity and tranquillity. And when Empedocles writes 'All the limbs of the gods were shaken' (fr. 31, quoted on p. 87), we need not take him to mean more than a great stirring-up of all four elements. Certainly the commentators say that Love and Strife exert their influence on 'the whole.' 'The whole (*to pan*) is split up into the elements by Strife, and fire is gathered together into one, and so is each of the other elements' (Aristotle, *Metaphysics* I.4, 985A25). Such language is typical. Strictly interpreted, it means that all the fire there is in the universe was gathered into one. But it can perfectly well be taken to mean only that there is a gathering of some fire, previously all mingled with other elements, into a single large mass. Certainly Empedocles spoke of the 'victory' (*kratein*) of Love and Strife in turn, and if Love's victory is the total harmony of the whole universe, this suggests that Strife's victory should also be all-embracing. But it is clear that this language is used to describe a number of partial victories of one or the other of the two forces (for example, in the birth and destruction of a single organism); and there is nothing in the fragments or testimonia that absolutely requires us to believe that *every bit* of the Sphairos is at some time overpowered by Strife.

It is likely, then, that Empedocles envisaged the first stages of the cosmos very much as Anaxagoras did. Somewhere in the primitive mixture of all things, a motive agent (Anaxagoras' Mind, Empedocles' Strife)

[9] See Guthrie, *HGP*, vol. II, p. 180. One who does not reject it is O'Brien: but disappointingly he relegates it to half a footnote (*Cycle*, p. 145, n. 3), and I do not understand how he would, or could, integrate it into the rest of his account of Empedocles. He makes the useful point that Plato's argument that no part of the elements is left outside the cosmos, in *Timaeus* 32C, may be directed against Empedocles.

started a disturbance. In Anaxagoras' theory, there is evidence for a beginning on a small scale, taking the form of a rotation, which caused a sorting of things by kind. Probably the same is true for Empedocles. At least we have clear evidence of a vortex at some stage in the gestation of the cosmos, in a fragment that must now be quoted:

> When Strife had reached the lowest depths of the vortex,
> and Love has come to be in the middle of the whirl,
> here all things begin to come together so as to be one,
> not all at once, but combining at will from different directions.
> And even as they mingled, mortal kinds in thousands issued forth.
> But many things stand unmixed, alternating with things mixed,
> as many as Strife still held aloft – for he had not blamelessly
> retired utterly from them to the outermost limits of the circle,
> but remained in part within and in part had gone out from the members.
> As much as he continually ran forth, so much there continually entered
> a gentle-hearted immortal stream of blameless Love.
> At once, what before had learned to be immortal grew mortal,
> and the hitherto unmixed was mixed, having changed course.
> And even as they mingled, mortal kinds in thousands issued forth,
> endowed with all manner of forms, a wonder to behold.[10] (fr. 35.3–17)

These lines have generated much controversy, particularly with regard to time and place, but we shall not find them too difficult so long as we do not try to fit them into a *spherical* picture of the world with a geocentric theory of motion, or a theory of *two* generations of mortal kinds. What is described here, I believe, is the kind of vortex exemplified by a whirlwind or tornado, in which the heavy, large objects are heaped at the bottom in the middle, and the lighter and smaller objects are carried upwards in a spiral and sprayed outwards at the top. Strife is the disturbing force, which takes things from their place of rest and separates them from each other according to their weight and size. Love is the force that dominates the still center of the vortex. The opening line, then, mentions the culmination of Strife's work of disruption in the Sphairos. Strife is represented as descending to the 'lowest depths of the vortex' – that is, as far down as the vortex is ever going to move. The word translated 'lowest' (*enertaton*) is from a root used especially of the underworld, and 'depths' (*benthos*) may suggest the bottom of the sea. The lowest point reached by the action of Strife is perhaps where the earth and the sea will be when the cosmos is fully formed. This is the time when Strife's domination in the region that

[10] The tenses in lines 3–5 are surprising, but I think O'Brien has made good sense of them, and I follow him (*Cycle*, pp. 108–13). Line 7 is repeated at line 16, and Diels therefore substituted another line in its position at 7. In line 15, I follow Rosemary Arundel Wright's reading in *Classical Review* 12 (1962).

will become the cosmos is most complete. Everything is in turmoil: the whirlwind hurls everything around.[11]

But the whirlwind has a still center, and at the bottom of the central axis, as the heavy elements accumulate, the still center begins to grow. We can think of the whirlwind moving gradually away from the bottom center, outwards and upwards. Lines 5 and 6 refer to this stage: there is a gradual accumulation of heavy elements that come to rest (actually because of friction, but Empedocles would not know that), and then a progressive quietening of the lighter elements, like dust settling, as the storm lifts.[12]

After the separative vortex has lifted from the still center, which is the earth, the elements begin to be mingled through Love, and mortal creatures are formed. We have more evidence about this process, and we shall return to it in the next section.

While mortal creatures are being born – and dying – on earth, the whirlwind still continues above and around the earth. This probably refers to the heavenly bodies; Aristotle mentions that they are held in position, according to Empedocles, by the speed of their rotation (*De caelo* 284A25). Strife itself is withdrawing to the 'outermost limits of the circle'; this is presumably said because the whirling motion is seen to be greatest in the stars near the horizon: the stars move less, the nearer they are to the pole at the center. Strife has not yet totally withdrawn, however. Love is advancing, but slowly, and there is still a lot of Strife around. Something more can be said now about the shape of Empedocles' cosmos, and its internal dynamics.

Empedocles says that the height from the earth to the sky – its elevation from ourselves – is less than the measurement in breadth, the sky being more extended in this respect, because the cosmos lies like an egg. (Aetius II.31.4)

The only reasonable way of interpreting this, as it seems to me, is to imagine the cosmos as an egg that is wide in comparison with its height, with the widest point quite near the bottom. The earth is situated on the central axis at the widest point. We learn from another text that the axis of the egg is tilted with respect to the earth: Empedocles, like Anaxagoras and

[11] For details about the vortex in this fragment, see Tigner, *Isis* 65 (1974).
O'Brien's discussion of the words 'the lowest depth' is most helpful (*Cycle*, pp. 113–17), but spoilt by his attempt to fit the fragment into a spherical picture. After a careful and conscientious collection of reasons for taking the phrase to mean 'the lowest part,' he switches finally to 'innermost depth'!

[12] So it is hard to know whether the formation of the earth is due to Strife or Love. If I am right in taking Love to be a force that brings things to rest together and Strife as the cause of the vortex, then Earth would seem to be a product of both forces, whereas the formation of the heavens owes more to Strife. See below, section 7.6.

others, uses this *ad hoc* device to explain why the orbits of the heavenly bodies are not parallel to the plane of the earth.[13]

The dynamics that produced such an egg-shaped cosmos are plausible enough. The whirl separated the elements so that air and fire rose up the axis to the top, where they were dispersed outwards, to fall in a downward spiral so as to form a wide dome over the heavy elements that assembled in the middle at the bottom. The earth was actually lifted from the bottom and is held in suspension by the whirl – so much is said by Aristotle, although some details of this are obscure. The air and fire of the heavens closed under the earth. The pressure of the heavy bodies at the center, we are told, squeezed the water out of the earth so that it lay on top, or else it was sweated out by heat; and the dome of the sky was filled with air exuded from the water (Aetius II.6.3; III.16.3).

Aristotle's description is in *De caelo* II.13, 295a13–22:

> Therefore all who hold that the world had a beginning say that the earth travelled to the middle. They then seek the reason of its remaining there, and some claim, as we have said, that its flatness and size are the cause; others agree with Empedocles that it is the excessive swiftness of the motion of the heaven as it swings round in a circle which prevents motion on the part of the earth. They compare it to the water in a cup, which in fact, when the cup is swung round in a circle, is prevented from falling by the same cause, although it often finds itself underneath the bronze and it is its nature to move downwards. (trans. Guthrie)

Tigner, *Isis* 65 (1974), argues that Aristotle has applied the wrong illustration to the problem of the position of the earth. The twirled ladle, he argues, is probably meant to illustrate some point about the motion of the heavenly bodies. The feature of the swirl that might explain the position of the earth is that a flat plate can be raised from the floor of the container and held in suspension in water that is swirled vigorously enough.

There remains a mystery that I cannot penetrate. We are told that Empedocles divided the sky into two 'hemispheres' – presumably that means that he distinguished two halves of the egg – and claimed that one is 'fiery' and the other, that of night, 'airy.' The sun that we see is a reflection, from the disk of the earth, of the whole fiery 'hemisphere': the earth, that is to say, acts like a burning-mirror, collecting the light of the whole sky and returning a narrow beam that we see when it throws a circle of light onto the shell of the sky. There are obviously many difficulties about

[13] Aetius (II.8.2) even gives a mechanical explanation of the tilting of the axis: 'the air yielded to the onrush of the sun.' I do not know what this means.

For another possible description of the Empedoclean cosmos, see Bollack, *Empédocle*, vol. III, pp. 253–302. It is possible, as Guthrie suggests (*HGP*, vol. II, p. 190) that Empedocles derived the idea of an egg-shaped cosmos from Orphic poems. See West, *The Orphic Poems*, index s.v. 'egg, cosmic.'

such a theory, even if this simplification of some confused texts is correct, but since I cannot see a way through them it seems pointless to carry the discussion further here.[14]

The dynamics of the vortex, and such descriptions of the motions of the elements in the cosmos that we have, suggest that Empedocles still holds to the linear model for cosmic dynamics; he has not yet arrived at the centrifocal pattern. He probably thought of the earth as flat.[15] The heavenly bodies whirl around the earth in circular orbits, but perhaps form an egg-shaped pattern as a whole, rather than a sphere.

7.4 *'Mortal kinds in thousands, a wonder to behold'*

Empedocles was the first writer to produce a 'theory of evolution' of which some fragments have survived. It was adopted by the Atomists with very little change, and became a vitally important part of the anti-Aristotelian cosmology. As a matter of fact, in spite of its imaginative brilliance, it turned out to be the weakest point in the Atomists' case. Ultimately, the best intellects of the ancient world would not be convinced that living forms could have grown by nature out of inanimate elements, and preferred to attribute them either to a creator god, or to an eternal, non-evolutionary nature of things. There is nothing surprising in that, of course; Darwin's theory of evolution was received, only a century ago, with suspicion, incredulity, and passionate rejection, for some of the same reasons that persuaded the Greeks and Romans to turn away from Empedocles, Democritus, and Lucretius. It is only since that time that we have been able to realize that Empedocles was not just indulging in fantasy, but was putting an essential item on the natural scientists' agenda.

The greatest interest focuses on the generation of animals, but a few scraps of information about the earlier growth of vegetable life have survived. Trees – perhaps greenery in general – grew from the earth before the sun 'was unfolded and spread around' (*sic* – the true sun is a 'hemisphere' of the heavens, whose reflection off the earth is the visible sun), and before day and night were distinguished. Owing to the harmony of the blend of elements in them, they 'combine the formula' of male and female. The heat in the earth made them grow.

[14] The texts are *Stromateis* ap. Eusebius, *Praep. evang.* I.8.10 = DK 31A30; Aetius II.11.2 = DK 31A51; Aetius II.20.13 = DK 31A56. For further discussion, see Guthrie, *HGP*, vol. II, pp. 190–9, Ferguson, *Phronesis* 9 (1964).

[15] R. A. Wright says (*Empedocles*, p. 200): 'From his explanation of eclipses, and the earth being held still by the rotation of the sky, it is clear that he envisaged it as spherical.' Empedocles had the right explanation of eclipses, but this does not entail that he thought the earth is spherical; a disk will do just as well. If Tigner's explanation of the vortex is right, it cannot explain the position of a spherical earth, and Wright does not offer any explanation.

In another place Aetius gives an account of the development of living creatures, by stages. This has been sadly mauled and split up by some of the modern commentators dominated by the dogma of the reversing cycle, which we shall discuss in the next section. In my view it is coherent and consecutive, although not without some puzzles.

Empedocles held that the first births of animals and plants were by no means whole, but disjointed, with parts that were not growing together. The second, when the parts did grow together, were like dream images. The third were of whole-natured things. The fourth were no longer from (?) similar things such as earth and water, but already through each other – some when the nutriment is condensed, others when the shapeliness of the females effectuates a stimulation of the spermatic movement. (Aetius V.19.5 = DK 31A72)

The last sentence is a terribly pompous piece of Greek, and I have tried to translate it into appropriate English.

Some of the four stages listed by Aetius can be illustrated by surviving fragments of Empedocles' poem. The first – a famous and haunting vignette – was aptly quoted recently by a writer in the *New York Times* at the beginning of an article on contemporary fears about recombinant genetic biology:

Many heads grew, neckless:
arms wandered bare, deprived of shoulders;
eyes strayed alone, in want of foreheads. (fr. 57)

This corresponds with Aetius' mention of 'parts that were not growing together'; their conjunction is described in these lines:

But when god mingled all the more with god
these things fell together as each of them chanced to fall,
and many others beyond these grew, in continual succession. (fr. 59)

Simplicius reports that this belongs, as we should expect, to the period when Love's influence begins to recombine the elements after their separation by Strife. The elements are referred to as gods in line 1, presumably because of their indestructibility. It is noteworthy that in spite of this appellation, and in spite of their being under the influence of Love, their meetings and minglings are ascribed to chance.

The second stage in Aetius' narrative is the production of whole creatures, not merely parts, but creatures that combine unmatched parts, like the figures of a dream or nightmare:

Many grew double faced and double chested;
manfaced oxlings, and again there were produced
oxheaded manforms – mixed creatures, part from men,
part of woman's form equipped with shadowed parts. (fr. 61)

Plutarch mentions also some 'hand-and-foot rollers,' and Aristotle, parodying, asks why there are no 'olive-faced grapelings.'[16] It is not clear whether these mixed-up animals are the result of the promiscuous desire of loose limbs for company, or are themselves born, fully misshapen, from the earth.

What is important is the idea of natural selection by survival of the fittest, not attested in Empedocles' own words, but plainly attributed to him by Aristotle. In a passage that we shall examine in more detail in volume 2, Aristotle outlines his opponents' case against his own theory of the teleology of nature. Perhaps after all, they say, everything in nature might be fully explained in terms of matter-in-motion:

> What, then, is to stop parts in nature too from being like this – the front teeth *of necessity* growing sharp and suitable for biting, and the back teeth broad and serviceable for chewing the food, not coming-to-be *for* this, but by coincidence? And similarly with the other parts on which the 'for something' [i.e. the final cause] seems to be present. So when all turned out just *as if* they had come-to-be for something, then the things, suitably constituted as an automatic outcome [i.e. just from matter-in-motion], survived; when not, they died, and die, as Empedocles says of the manfaced oxlings. (*Physics* II.8, 198b24–32; trans. W. Charlton, adapted)

Manfaced oxlings were ill adapted for life and that is why we never meet one. This became an essential part of the Atomists' case; although the writings of both Democritus and Epicurus on the subject have perished, Lucretius' use of it is enough to prove its importance (*De rerum natura* V.855). It was part of the Atomists' thesis that the world is all the result of accident, and so they had to explain why, in so many respects, it does not look like an accident. Their answer was that what we now see around us is selected, by its fitness to survive, from a palaeontological random miscellany. Evidently, this thought had already been worked out by Empedocles.

To return to Aetius' list, we now come to the third step. 'Whole-natured things' is Aetius' phrase, and we have the same word in a fragment of the poem:

> Next, how men's and much-lamenting women's
> night-born offspring were produced by separating fire,
> hear this – my tale is not aimless or unknowing.
> Whole-natured forms arose at first from the earth,
> possessing a measure both of water and of heat.
> They, then, were sent forth by fire, wishing to reach its like,

[16] Plutarch, *Adv. Coloten* 28, p. 1123b = DK 31B60. Aristotle, *Physics*, II.8, 199b11, in DK 31B62.

> creatures displaying no lovely form of limbs
> nor voice and member proper to men. (fr. 62)

Although this fragment has been much debated, certain points in it seem sufficiently clear. Empedocles asserts that *before* men and women produced children through sexual reproduction, there was a time when the *earth* produced beings that were somewhat like humans but lacked certain human 'limbs' such as sexual organs. The opening words of line 6 (τοὺς μέν) suggest that he continued by spelling out the other half of the antithesis: *these* creatures had no means of reproducing themselves – but others emerged from the earth at this early period who were capable of 'mingling in Love' and so of continuing the human species. The fragment fits well enough into the pattern suggested by the passage of Aristotle translated on p. 19 (*Metaphysics* I.3, 983b6–984a10). The world in which we live is populated by the children of those who once upon a time *happened* to spring from the earth equipped with the means to perpetuate their kind.[17]

Plato makes fun of this idea in the *Symposium*, in the speech in praise of Love that he gives to Aristophanes. Aristophanes' fantasy is that men and

[17] The first problem is whether the 'whole-natured forms' are contrasted (a) with the separate limbs of earlier stages, or (b) with the contrasted male and female forms of the present day. In other words, which of these is the emphasis of line (4): (a) whole-natured forms first arose from the *earth* (not sexually), or (b) the first (near-humans) to arise from the earth were *whole-natured* forms (not sexually differentiated)? It seems to me that in the fragment itself (a) is a most unnatural interpretation. There is no trace in the fragment of the contrast with earlier stages. So I prefer (b), with most commentators; for the contrary case, see Bollack, *Empédocle*, vol. III, pp. 430–1.

The second problem is how to fit the fragments into the cosmogony. Does the mention of 'fire wishing to reach its like' not suggest the work of *Strife*, and are not sexually differentiated forms *more* separated (by Strife) than whole-natured forms? Guthrie (*HGP*, vol. II, pp. 206–7) and O'Brien (*Cycle*, pp. 199ff.) answer 'Yes' to both questions, and so place this stage of growth in the *opposite phase* of the cycle to the formation of isolated and unfitted limbs. Bollack answers 'No' to both questions. I answer 'No' to the second, and 'Perhaps' to the first.

On the second question, it seems to me that since Love as a cosmic force starts as a sexual metaphor, it is no surprise that Empedocles thinks of sexual union as showing Love's influence in a very high degree. There is no reason to think that a creature without sexual differentiation would represent a higher stage, even though the Sphairos (Love's supreme achievement) is explicitly described as lacking human limbs (fr. 29). There are less exalted beings that may have been in Empedocles' mind – for example, an embryo before limbs are distinguishable (Aristotle says 'the whole-natured was seed': *Physics* II.8, 199b7). The main feature of these whole-natured forms, I take it, was that they could not reproduce themselves, and I doubt if that argues for a high degree of Love's power, as compared with sexual pairs.

On the first question, see below, section 7.6. It seems to me that such a creative movement, even if it tends in the direction of the isolation of elements from each other, might have been thought of as the work of Love. Or else the separation of fire is the work of Strife, still lingering in the cosmos after all; but it is not necessarily a sign of *increasing* Strife.

women were once upon a time combined into 'whole forms,' but were later split in half by the gods to prevent them from being too powerful. As a result, human beings of the present day are yearning to be reunited with their other half; and that is the nature of human love. Since this is plainly a comic invention, it can hardly serve as reliable evidence for the role of the 'whole-natured forms' in Empedocles' theory. But it may be worth noting that the stage in which humans were not sexually divided precedes the stage of sexual division in Aristophanes, and that Love is mentioned in connection with the latter phase. This is not false to Empedocles' idea of the sequence of events, I believe; sexual union is the paradigm case of Love's work in the cosmos.

This, then, is the imaginative story that may be called Empedocles' theory of 'evolution.' It should be pointed out that it – and its successor in the later theories of Democritus and Epicurus – differ in one, all-important way from the nineteenth-century theory of evolution. The ancient theory did indeed claim that the world evolved, in the sense that there was a gradual progression in time from isolated elements to organisms, and they understood that the relevant kind of fitness was fitness to produce offspring. But they had no idea of the evolution of species. They knew nothing of genes, still less of mutations. They had some idea of variation within a species, but did not arrive at the notion of adaptive change in the character of a population. Their theory was not that the fittest individuals in a species survived to transmit their inheritable characteristics to the next generation, but rather that only the fittest species survived. The materialists as well as the Aristotelians thought of the surviving species as fixed entities.

7.5 *The question of the cosmic cycle*

> In turn they prevail, as the cycle goes round,
> and they pass away into each other, and grow in their portion of time.
> There are just these things; running through each other they become
> men and the tribes of other animals,
> at one time coming together into one ordered whole (*kosmos*) through
> Love,
> then again carried each apart, through the hatred of Strife,
> until they grow to be one entirely and thus pass into defeat.[18]
> Thus, in that one has learned to grow out of many,
> and again, when one grows apart, many come into being,

[18] This is a difficult line; even the text is in doubt. Perhaps it repeats the sense of line 5, as Wright, *Empedocles*, suggests, in order to indicate the cyclical nature of the process (although Wright would not agree with me about what that sense is).

> in this way they come-to-be, and their life is not for ever;
> but in that they never cease their continual exchange,
> in that way they *are* for ever, unchanging in their cycle. (fr. 26)

Empedocles uses pronouns for subjects in this passage, and that is no doubt a careful choice. He is writing about the elements and their formation of compounds, but he does not wish always to distinguish between the elements and the compounds. The elements are the compounds; the compounds are nothing but the elements. These things 'come-to-be' – but are also eternal.

Nobody doubts that the passage applied to the cycle of birth and death of plants and animals – 'mortal things.' The message is the same as that of fr. 17, quoted above on pp. 83–4: indeed, the last five lines are identical with lines 9–14 of fr. 17. To some, however, the two fragments have suggested that the subject is not only the cycle of birth, death, and rebirth of mortal things, but also a similar cycle in the history of the cosmos itself.[19] They have suggested that the cosmos, according to Empedocles, passes from a state of unity, in which the elements are so mixed together as to be indistinguishable, to a state of plurality in which the elements are not mixed with each other at all; and then back – or onward – to unity again. Unity, according to this scheme, is the state of the cosmos when Love completely prevails (the Sphairos), and the divided elements are the summit of Strife's achievement. Between the alternating triumphs of Love and Strife are two intermediate periods: one in which Strife encroaches on the unity produced by Love and breaks it up, and the other when Love progressively reunites the separate elements into one.

In fr. 17 Empedocles wrote:

> Two-faced (δοίη) is the birth of mortal things, two-faced their passing:
> the one is born – and destroyed – by the coming together of all things:
> the other, fostered by their growth again apart, is also thus dispersed.[20]

The adjective δοίη does not make it clear what kind of 'two-ness' is meant. I have translated it 'two-faced'; it has also been taken to mean

[19] The reversing cycle goes back as far as Zeller, although he thought Aristotle was wrong if he thought our world is the offspring of Strife's increasing power and not of Love's. It was adopted in such standard works as Burnet's *Early Greek Philosophy* (1892), Kirk and Raven, *The Presocratic Philosophers* (1957), Guthrie's *HGP*, vol. II (1965). It was assumed as true, and worked out in much detail, in O'Brien's *Empedocles' Cosmic Cycle* (1969). In spite of much criticism, it still appears in Barnes' *The Presocratic Philosophers* (1979) and in Wright's edition of the fragments (1981). It has been dropped in the latest edition of Kirk and Raven (2nd ed. by Kirk, Raven, and Schofield, 1983).

[20] The text is corrupt at the end of the third line quoted, and both the verbs 'fostered' and 'dispersed' are emendations. Earlier attempts to emend the text are well described by O'Brien, *Cycle*, pp. 164–8, and I follow the reading he prefers – although I differ from him in my interpretation.

'double,' and this is interpreted to mean that there are two processes of birth and two of death. In each case, one process is brought about by the increasing power of Love in the cosmos, and the other by the increasing power of Strife. These processes are then allocated to the two intermediate periods in the cosmic cycle. There is a sentence in Aristotle's *De generatione et corruptione* (II.7, 334a5) that provides the next step: 'At the same time [Empedocles] asserts that the cosmos is in a similar state now, in the time of Strife's influence, and previously, in the time of Love's influence.' So it seems that we are 'now' in a period when Strife is in power. The structure of the reversing cosmic cycle is now complete: we have

(1) The Sphairos, in which there is no movement;
(2) Encroachment of Strife on the Sphairos, bringing about a separation of the elements from one another – the first birth and death of mortal things, in which we now live;
(3) A state of total separation of the elements, whether momentary or of some duration;
(4) Increase of Love's influence, bringing about renewed mixing of the elements – the second birth and death of mortal things;
(5) = (1)

The result of this is little short of a catastrophe. Virtually all the evidence that survives concerning Empedocles' beautiful theory of the origin and development of life on earth plainly attributes it to the force of Love. Yet our world, according to the evidence of Aristotle, is in the time of Strife's power. The stages of 'evolution,' it is claimed by those who believe in the reversing cycle, occur in the reverse order in the two halves of the cycle. So the neckless heads, the manfaced oxlings, and other weird creatures, which belong in the prehistory of Love's world, are a part of the future of our world. In our part of the cycle, the survival of well-adapted creatures is irrelevant: they are doomed in any case to give way to 'eyes in want of foreheads' and other lonely limbs. Aristotle's criticism of the

---

These three lines were almost certainly written with Homer, *Iliad* VI.146ff. in mind:

> As (οἵη) is the generation of leaves, so (τοίη) is that of humanity.
> The wind scattered the leaves on the ground, but the live timber
> burgeons with leaves again in the season of spring returning.
> So one generation of men will grow while another dies. (trans. Lattimore)

Empedocles' δοίη ... δοίη exactly match Homer's οἵη ... τοίη in their position in the line. See Long, 'Empedocles' Cosmic Cycle in the Sixties', in Mourelatos, *The Pre-Socratics*, p. 401, n.9., for this parallel, which he attributes to A. H. Griffiths.

My own view of these lines appears to be thoroughly eclectic. I agree with Hölscher, *Hermes* 93 (1965), about the interpretation but not the grammar, and with Mansfeld, *Phronesis* 17 (1972), about the grammar but not the interpretation. Mansfeld's view, that the coming together of the elements both suggests the idea of birth and (as soon as it is properly understood) abolishes the idea, appeals to me as an intellectual construction, but finally it seems to make it too much of a 'meta-statement' to suit the context.

theory in *Physics* II.8, which assumes that it is meant to be an explanation of our world, is wholly beside the point. And the one connected account of the stages of development, reported by Aetius and quoted above on p. 95, has to be dismissed as a jumbled mixture of stages from different world periods.

In recent years, many Empedoclean scholars have come to think that the orthodoxy of the first half of the twentieth century was wrong, and the reversing cycle must be jettisoned.[21] But there is no agreement on an alternative interpretation. Some find the evidence for a recurring cycle on a cosmic scale too compelling to be rejected altogether. They seek to rescue the theory from the fantasy of a reversed 'evolution' of life by a different evaluation of the evidence for the periods of the cycle. The period in which Strife disrupts the Sphairos – period (2) – is interpreted as one that produces no life at all, but only the separation of the elements so as to form the main frame of the cosmos: separate masses of earth, rivers and oceans, air, and the fiery heavens. The development of life occurs only when Love begins to mix the separated elements, in period (4). Aristotle's remark that Empedocles' cosmos is 'now, in the time of Strife's influence, similar to what it was previously in the time of Love's influence' is either dismissed as being mistaken, or interpreted as telling us something about the natural motions of the elements during their separation into world masses and their mingling in the Sphairos.

More radically, some have argued that the cosmic cycle is nothing but a chimera – a creature put together by Aristotle (even by Plato) out of ideas torn from their context.[22] All of the fragments that mention or imply cycles refer to the ordinary cycles of mortal nature. The four elements combine to form a compound, and so the compound is born; the compound perishes and disintegrates, and so particular separate bits of the elements are born; they form another compound, and so on for ever. So Empedocles, like others of the Presocratics, would join the tradition of Hesiod's *Theogony*, in which there is a birth, followed by eternal life. This interpretation is supported by a very detailed analysis of Aristotle's testimony, designed to show how his mistaken impression of Empedocles' theory arose.

Agreement is far off. Two recent accounts of Empedocles, one of them a

[21] The 'reversing cycle,' which I have called 'orthodox,' was criticized by von Arnim, 'Die Weltperioden bei Empedokles,' in *Festschrift Theodor Gomperz dargebracht*, in 1902. Recent critics include Bollack in *Empédocle* (1965), Hölscher in *Anfängliches Fragen* (originally *Hermes* 96 (1965)), Solmsen in *Phronesis* 10 (1965), Mansfeld in *Phronesis* 17 (1972), Long, 'Cycle' (1974), and Van der Ben, *The Proem of Empedocles' Peri Physios* (1975).

[22] Hölscher, *Anfängliches Fragen*, and Van der Ben, *The Proem*.

full and scholarly study of all the fragments and testimonia, have returned to the reversing cycle.[23] It is my view that the implausibilities of this scheme are too great to be acceptable; but it is hard to be sure which of the two alternatives is right. Fortunately it is not of great importance for the theme of this book which of them is preferred. What is of importance is that we may attribute to Empedocles a linear theory of the growth of our cosmos, not a reversing one. The enduring achievement of Empedocles was his theory of the elements, their formation of compounds and dissolution into separate elements again through the agency of opposed forces of attraction and repulsion, and the progressive formation of more complex organisms from simpler components. These were essential steps towards the atomic theory.

7.6 *Appendix: the action of Love and Strife*

In the extant fragments of Empedocles, the joining of *different* things to each other is attributed explicitly to Love, designated by one or other of its names, in fragments 20.1–3; 21.8–13; 22.5; 35.12–17; 71.2–4; 73.1–2; 96.1–4; 98.1–3. There are no fragments in which the attraction of the elements to themselves is explicitly attributed to Love by name.

On the other hand certain texts are said to have this implication: arguments can be found in Carl Werner Müller, *Gleiches zu Gleichem*, pp. 27–39, Jean Bollack, *Empédocle*, vol. 1, pp. 48–52; Michael C. Stokes, *One and Many in Presocratic Philosophy*, pp. 168–72, and A. A. Long, 'Empedocles' Cosmic Cycle in the Sixties', pp. 412–18. I find the arguments tempting, but finally unconvincing. None of them has a cogent reply to the arguments of Harold F. Cherniss, *Aristotle's Criticism of Presocratic Philosophy*, p. 190, n. 193.

A lot of weight is placed on fragments 21 and 22. Thus Long claims that in fragment 21 Pausanias is invited to consider 'the sun, the air, the rain, and the earth ... the clear implication of this text is that the sun, air, earth, and water – the main cosmic masses which correspond with the four elements – each consist now of aggregates of like elements put *together by Love* [Long's italics].' But I find this implication by no means clear. What is said is that 'in Strife they are all different in form and divided, but in Love they come together and *desire one another*,' and the reason given for this is that from them come plants, animals, men and women, and gods. The *only* work attributed here to Love is the mingling of the elements in compounds. On the other hand, I concede that it is not said that the *cosmic masses* are the work of Strife. What is said is that the

[23] Barnes, *The Presocratic Philosophers*, and Wright, *Empedocles*.

four elements, under Strife, tend not to unite with each other (the sun is evidently mentioned, along with air, rain, and earth, as an *element*, not as a cosmic mass, because in lines 9ff. trees etc. are said to be made of these four). It is still possible that the cosmic masses are held *in relation to one another* in the cosmos by Love.

Fragment 22 is very difficult. O'Brien (*Cycles*, pp. 305–13) analyzes it well, and I believe he is right about its structure. In the first three lines we are told that the parts of the four elements are 'at one' (ἄρθμια) with themselves, even when they are scattered (ἀποπλαχθέντα) in mortal compounds; in the last six lines, by contrast, we are told that those (parts?) of the elements that are adapted for mixing, and love *each other* through the influence of Aphrodite, are also most at variance with each other and unadaptable to mixing, through the power of Strife. 'In the second half,' Long observes, 'joining together and separation are explicitly attributed to Love and Strife respectively. Is there any reason why these powers should perform a different role in the first half?' Well, nothing is said about the use of the word ἄρθμια, which *is* explicitly attached to Love in fr. 17.23. But to press that point here is to beg the question: what we want to know is precisely whether ἄρθμια ἔργα *between likes* are the work of Love. What is striking about fragment 22 is the fact that it *does* mention Love and Strife in connection with the relation of the elements *to each other*, and *does not* mention them in connection with their relation to themselves.

In fragment 62.6 we have a mention of fire 'wishing to reach its like.' If every instance of a verb meaning 'to want or desire' is to be attributed to the force of Love, then this is a case of like-to-like attraction caused by Love. But it is a shaky foundation to build on, since this movement of fire takes place in a situation in which it is mixed with earth, and it is a movement that contributes to the generation of living things. Even if we can conclude that like-to-like attraction is sometimes the work of Love, which is doubtful, we have no right to conclude that *all* such attraction is the work of Love.

The evidence of Plutarch (fr. 27, from *De facie*) described on pp. 88–9, seems to me to suggest that he thought the assembly of the elements in single, separate masses was the work of Strife. Although there is a puzzle as to why he says the sun, the earth, and the sea are indiscernible when the elements are separated by Strife, as Long points out, it may just as well be due to the idea that these items in the physical world are actually mixtures, and are held in this position in the cosmos by Love, as to the scattering of the elements into isolated particles, as Long claims. Actually there is no evidence of a state of isolated particles: μονάδες in Plutarch's description clearly means no more than 'non-mixtures.'

In fragments 17 and 26, Empedocles makes much of the alternation between one and many. The birth of one is the death of many, and vice versa. How does he relate this alternation to Love and Strife? In principle, he could either attribute birth always to Love and death always to Strife, or ascribe unity to Love and plurality to Strife. It seems clear to me that he chooses the latter option, and that 'unity' in these fragments means a conjunction of different elements, and 'plurality' a separation of the elements. It may be claimed that these fragments are concerned only with the generation and destruction of organic compounds, and so nothing can be learned from them about the generation and destruction of simple homogeneous masses. Perhaps so; but it is an arbitrary restriction of meaning.

> now, it grows to be many again, out of one:
> Fire and Water and Earth and the vast height of Air (fr. 17.17–18)

The production of many is the work of Strife; and nothing here says that the many produced by Strife must be isolated particles, rather than just four masses.

8 Later Eleatic critics

The crucial importance of Parmenides of Elea in the history of cosmology has been explained. We must now turn to his followers, who reinforced his arguments with contributions of their own. Some of these played an extraordinary and fascinating part in the development of Atomism, and their authors have an assured place in history. It happens that these connections have been fairly thoroughly explored in recent literature, especially in the English language, and this chapter will accordingly be relatively brief.

We hear of two followers of Parmenides, each about one generation younger than he was. Zeno came from Parmenides' home town of Elea, Melissus from the Aegean island of Samos, the birthplace of Pythagoras. Their work can be dated about the middle of the fifth century. There is no doubt about their allegiance. For Zeno, we have Plato's authority: his dialogue called after Parmenides presents Parmenides, as a man of sixty-five, with Zeno as a companion then aged about forty, in conversation with Socrates, at the time still very young. The occasion itself is probably fictitious, but the relative ages are likely to be about right, and when Plato has Zeno say that his book was in support of Parmenides' argument, he may be believed. Against those who ridiculed Parmenides' thesis 'One thing is,' Zeno is made to say, he wrote to show that still more ridiculous things follow from the contrary thesis, 'Many things are' (*Parmenides* 128b–c).[1] Melissus' argument, as reported in one form by Simplicius and in another by a pseudo-Aristotelian monograph about him, is a parallel, in prose, to Parmenides' Way of Truth, differing from Parmenides in the details of the argument, but using similar premisses and reaching the same conclusion.

Melissus' book *On Nature* or *On What Is* was probably written later than the book by Zeno referred to by Plato.[2] It is my belief that both are to be dated after Anaxagoras and Empedocles but before Leucippus and Democritus; some historians, agreeing that there is a link between the work of Anaxagoras and Zeno, and again between Melissus and Leu-

[1] Solmsen, *Phronesis* 16 (1971), attempts to throw doubt on the historicity of Plato's statement, but I am not convinced. He is answered by Vlastos, *JHS* 95 (1975).

[2] He may have written more than one: the Suda says he wrote *Contentions, Exegesis of Empedocles' Works, Against the Philosophers*, and *On Nature* (DK 29A2).

cippus, claim that the link in each case works in the opposite direction. I shall not argue in detail about the chronology here; if the connections that I shall suggest are plausible, that in itself will lend some support to the chronology.[3]

8.1 *The paradoxes of plurality*

Parmenides, as we have seen, laid down a challenge to anyone following him who proposed to write about nature. Nature involves change: change involves plurality. He attempted to show that neither plurality nor change could be consistently and rationally described, since both involved 'what *is not*,' and that cannot be part of any rational discourse. The challenge was to find a way to write about change and plurality without fatally introducing 'what *is not*,' or to prove Parmenides' premiss false by showing that it is after all possible rationally to talk about what *is not*. We have seen, in chapter 5, that the first way was attempted by Anaxagoras and Empedocles.

It is reasonable to think that Zeno was giving an answer to these two, although his arguments were apparently directed at a wider target – namely, at all the ways he could think of in which what *is* might be supposed to be divided up into a plurality.[4]

Perhaps the most famous of his arguments is the so-called 'Achilles' (Aristotle, *Physics* VI.9, 239b14ff.). The scenario is a race between Achilles and the tortoise, in which the tortoise, being much slower, is given a long start. Suppose the race starts at time $t(1)$, with Achilles situated at point P(1) and the tortoise further along the track at point P(2). The race begins. When Achilles reaches P(2) at time $t(2)$, the tortoise has advanced a little from P(2) to point P(3). When Achilles reaches P(3), at time $t(3)$, the tortoise has advanced a little, from P(3) to P(4). When Achilles reaches P(n) at time $t(n)$, the tortoise has advanced to P($n+1$). There is no $t(n+m)$ such that at that time Achilles and the tortoise will be at the same point. In other words, Achilles can never catch up with the tortoise.

This lovely paradox is obviously not part of the Form Book for Heroic Age punters: it is a parable of some kind. One interpretation might be this: if we suppose that a distance is divisible at points like P(1), P(2) etc., and that things may move over this distance at different speeds, we get the

[3] See chapter 6, note 1.

[4] I have written about Zeno in this same context in my *Two Studies in the Greek Atomists*, pp. 57–79, reprinted in part in Mourelatos, *The Pre-Socratics*. I shall try to keep repetition of this to the minimum in this chapter. The best account of Zeno known to me is Vlastos, 'Zeno of Elea,' in Edwards, *Encyclopedia of Philosophy*, vol. VIII. Recent contributions to the subject have been made by Barnes, *The Presocratic Philosophers*, vol. I, chs. 10–13, and Sorabji, *Time, Creation, and the Continuum*, ch. 21.

paradoxical result that 'faster' does not mean what we know it means. We must reject either the premiss that distance is so divisible, or the premiss that things may move at different speeds.

Aristotle points out that this paradox is 'the same' as a simpler one called 'the Dichotomy,' which he also describes (*Physics* VI.2, 233a21). This takes as its datum that a distance AB to be traversed by a moving object (the scene is usually thought of as a stadium) may be divided progressively in halves, first at C, then CB in half at D, DB at E, and so on.

....

A C D E B

Before the mover can go from A to B it must first reach C; before it can go from C to B it must first reach D; and so on *ad infinitum*. There is no *last* move in this series, such that having completed it the mover will be at B. If we reject this conclusion as absurd, we must reject either the premiss that the distance can be divided by progressive dichotomy or the premiss that a thing may move by passing through one part after another.

Both of these paradoxes, it will be seen, are so composed as to bring into question either a premiss about the divisibility of a length of territory, or one about motion. They are labeled 'arguments about motion' by Aristotle, and so they have often been interpreted, in ancient and modern times, as being efforts to prove the impossibility of motion. There are two difficulties about this interpretation. One is that Plato tells us that Zeno argued against plurality, not against motion. The second and more serious one is that both have to concede the possibility of the division of the distance in order to refute the possibility of motion: it is only if the stadium is *divisible* that we can show that it is impossible to reach its end.

It is possible, however, to work both of these arguments into a scheme of argument against plurality, if we assume that the movement of thought, when one thinks of the notion of a distance, qualifies as a motion over a distance. If to think of a distance is necessarily to think first of one part, then the next part, and so on, then the thinker, like the runner in the stadium, can never have a *last* move to make, by completing which he will reach his goal. This might be construed as showing that no object can successfully be thought of as a whole if its extension is divisible into parts.

But undoubtedly the most important argument of Zeno for this history is one that is reported in Simplicius' commentary on Aristotle's *Physics*. It has to be reconstructed from partial citations, but there is general agreement about its structure.[5] The conclusion of the argument is given by Sim-

[5] Zeno frs. 1 and 2. See my *Two Studies*, pp. 64–9; Vlastos, in *Encyclopedia of Philosophy*, vol. XIII, pp. 369–71, or Kirk, Raven, and Schofield, *The Presocratic Philosophers*, 2nd ed., pp. 265–9, for more detail.

plicius in these words, which have a good claim to be Zeno's own:

> If there are many things, they must be both small and large – so small as not to have any magnitude, so large as to be infinite. (Simplicius, *Physics* 141.6–8)

The argument for the first half of this conclusion is given only a summary account by Simplicius: 'each of the many things is the same as itself, and one.' It is not at all clear what this means. The interpretation I prefer is that it is derived from analysis of the concept 'many things': 'many' means 'one and one and one...' and thus if 'many' is to be intelligible the 'ones' that compose it must themselves be determinate and irreducible. If a 'one' were itself a 'many,' then the concept of many would have to be defined in terms of itself. But since each of the many must be a one that is not capable of internal differentiation but 'the same as itself,' it must have no magnitude, Zeno concludes. This apparently assumes that to have magnitude is to have one part distinguishable from another part, and thus to be in some sense not a one but a many.

Zeno next attacks the conclusion of this argument by showing that such a one, without magnitude, could not even exist:

> For if it were added, he says, to something else that *is*, it would make it no larger, since – it being of no magnitude, but added – the thing could not advance at all towards magnitude. And thus that which was added would of course be nothing. If indeed the other becomes no smaller when subtraction is made and again will not grow when addition is made, plainly what was added, and again what was subtracted, was nothing after all. (Simplicius, *Physics* 139.11–15)

This clumsy and archaic piece of prose presents some curious reasoning. It is hard to see how the initial assumption differs from the conclusion. The reader is supposed to agree with the assumption, stated in the last sentence, that if some X, added to or subtracted from an existent Y, makes no difference to the size of Y, the X 'was nothing after all.' Then it is claimed that the one, lacking magnitude, that was the product of the last argument qualifies as such an X, and thus it is concluded that such a one must be nothing. Is one to conclude that Zeno supposed it to be easier to believe that an existent thing must increase or diminish the size of whatever it is added to or subtracted from, than that an existent thing must itself have some size? It is quite unclear how he would respond to the objection that one may gain knowledge (say), or change color, without increasing in size. But perhaps his statements were meant to have a more restricted application than appears at first sight. The argument is directed against pluralist cosmologies: the 'many' under attack are the physical items postulated by those who offered accounts of the natural world. Zeno's pluralist opponents would surely be ready to concede that the physical world must be composed of parts that contribute to the bulk of the whole, and that a thing that makes no contribution does not qualify as a part of the physical world. And that is all the concession he needed. If this attack was on physical theories of a changing world made of plural material elements, he could afford to ignore non-material items.

His argument then continues with an objection to the thesis that the world is composed of plural material items that are not disqualified on the grounds just discussed:

> If it *is*, each [*sc.* of the many] must have some magnitude and bulk, and one part of it must stand distinct from the other. And concerning the outstanding part, the same story: for it too will have magnitude, and a part of it will stand distinct. Well, to say this once is like saying it always: for no such part of it will be the last, and not related as one part to another part.
>
> Thus, if there are many, they must be both small and large – so small as to have no magnitude, so large as to be infinite. (Simplicius, *Physics* 141.2–8)

The assumption Zeno needs for this argument may be put in this way: if a thing is qualified to be a part of the physical whole, then it must itself have parts of the same kind. And 'to say this once is like saying it always': in other words, if the physical world is divisible, it is divisible *ad infinitum*.

But why is this an objection to pluralism? The sting is in the tail Zeno gives to the argument: he claims that to show that such a being is infinitely divisible into parts is to show that it is 'so large as to be infinite.'

Different readings of this argument are possible, and there have been different interpretations of Zeno's conclusion and of the error (if it is one) in his reasoning. The procedure may be like that of the Dichotomy paradox described above. In that case it could be paraphrased thus. Within any bulky object, two parts may be distinguished, such that one 'stands distinct from' the other – e.g. as a limb does from the trunk. But the same is true of the part that plays the role of the limb – and so on for ever. If that is the right reading, the point may lie either in the (false) conclusion that the object so divided into parts increases in size beyond any finite magnitude, or in the (true) conclusion that one will never exhaust the object by such divisions and so never reach an ultimate, undivided part constituting its boundary.

Alternatively – and I now think this is more likely – the division may proceed by dividing not one of the two parts at each step, but both of them. The objection would then be that any bulky object must contain infinitely numerous parts, each having some finite size, and thus its size must be infinite. The fallacy in this has been well set out by Jonathan Barnes, in a way that convinces me: to summarize roughly, Zeno has not proved that his object consists of an infinite set of finite parts, but rather that it may be regarded as consisting of an infinite series of *finite* sets of finite parts.[6] But the fallacious conclusion is a plausible one that proved acceptable to many later philosophers.

Whatever may be the right interpretation of this argument, it depends on the premiss 'to say this once is like saying it always.' One way to escape from the awkward consequences of Zeno's arguments was to deny that premiss with regard to the division of the physical world. This was precisely the way chosen by the atomists Leucippus and Democritus. They asserted that not everything that has physical bulk contains parts such that 'one part must stand distinct from the other.' This is true of large- and medium-scale physical objects that we perceive with our senses, but it is not true of the ultimate components of these objects: the atoms.

It must be emphasized that atoms were postulated in order to meet the Eleatic arguments. There is no good evidence of any atomic theory before the time of Zeno.[7] Moreover, it seems unlikely that Leucippus and Democritus raised the claim that matter exists in the form of indivisible particles for reasons other than those of defense against the Eleatics. This is confirmed by Aristotle's statement in the *Physics* (1.3, 187a1–3) that some, meaning Leucippus and Democritus, 'gave in to the argument from Dichotomy by positing atomic magnitudes.'[8] We may believe that Leucippus and Democritus did not, so to speak, find atoms given them, either by tradition or by the demands of other problems in their physics, and then observe that these atoms would serve to fend off the Eleatic challenge against pluralism. They *invented* atoms as a defense against the Eleatics.

We shall follow up this theme in chapter 9, and observe what properties were thought necessary for the atoms to have if they were to serve this purpose.

8.2 *Melissus*

Three of the arguments that we find formulated for the first time in the fragments of Melissus are especially relevant to the formation of Atomism. There is no proof that Melissus invented them, or that Leucippus

[6] Barnes, *The Presocratic Philosophers*, vol. 1, pp. 251–2.

[7] For a criticism of theories of early atomism, and bibliography on the subject, see my *Two Studies*, Study 1, ch. 3.

[8] The ancient commentators on the *Physics* took this to refer to the Platonist Xenocrates, but their view was refuted by Cherniss, *Aristotle's Criticism*, p. 75, n. 303.

and Democritus read or heard them; but it is at least plausible that this is the case.

The first is an argument leading to the conclusion that what *is* is infinite.[9] 'Just as it *is* always,' Melissus writes (fr. 3), 'so it must always be infinite in magnitude too.' 'Nothing having a beginning and end is either eternal or infinite' (fr. 4). To understand the argument, we have to refer to his argument for the eternity of what *is*, since fr. 3, on the most likely interpretation, states that there is some kind of correspondence between the argument for eternity and that for infinity.

There always was what was, and always will be. For if it came-to-be, before it came-to-be it had to be nothing. Now, if it were nothing – in no way could anything come-to-be from nothing. (fr. 1, from Simplicius, *Physics* 162.25–7)

Since it did not come-to-be, it *is* and always *was* and always *will be*, and has no beginning nor end but is infinite. If it came-to-be, it would have a beginning (for it would have begun coming-to-be at some time) and an end (for it would have ended coming-to-be at some time). But since it did not begin and did not end, it always was and always will be and has no beginning nor end, for it is impossible that it should *be* always, if it *is* not wholly. (fr. 2, from Simplicius, *Physics* 29.22–8)

By substituting place-values for time-values, one can construct an argument for the spatial infinity of what *is*. If what *is* begins here, then it can only begin from what *is not*, and that is impossible.

The second argument, one that connects the possibility of motion with the existence of void, is embedded in a relatively long connected demonstration of the properties of what *is*, according to the principles of Parmenides. Since connected arguments of any length are rare among Presocratic fragments, it is worth quoting in full:

(1) In this way, then, it is eternal, and infinite, and one, and all alike.

(2) And it would not perish [*v.l.* lose anything], nor become greater, nor be changed in order, nor does it feel pain or anguish. For if it suffered any of these, it would no longer be one. For if it alters, it must be the case that what *is* is not alike, but that the former being perishes, and the non-being comes-to-be. If, then, by a single hair in ten thousand years it becomes altered, it will perish as a whole in the whole of time.

(3) But neither is it possible that it should be changed in order. For the order (*kosmos*) previously in being does not perish, nor does the order not in being come-to-be. If indeed nothing be added nor lost nor altered, how could any of the things that *are* be changed in order? For if anything became altered, *then* it would be changed in order too.

[9] There is controversy about the argument, of course. I find myself in general agreement with Barnes, *The Presocratic Philosophers*, vol. 1, pp. 200–4, and will therefore refrain from repeating the reasoning.

(4) Neither does it feel pain; for it could not all be in pain, since nothing could always *be*, feeling pain. Nor does it have equal power with the healthy. It would not be all alike, if it felt pain; for if something were lost or were added, *then* it would feel pain, and would no longer be alike.

(5) Nor could the healthy feel pain, since the healthy, and what *is*, would perish and what *is not* would come-to-be.

(6) And concerning anguish, the same story as with pain.

(7) And nothing is empty. For what is empty is nothing. Now, what is nothing could not *be*. Nor does it move. For it cannot yield anywhere: it is full. If it were empty, it would yield into the empty; there being nothing empty, it has no place to yield.

(8) It could not be dense and rare. For it is not possible that the rare should be full, like the dense: the rare thereby at once becomes more empty than the dense.

(9) This is the judgement that one must make of the full and the not-full: if it yields at all or is receptive, it is not-full; if it neither yields nor receives, it is full.

(10) Now, it must be full, if there is no empty. Now, if it is full, it does not move.

(Melissus fr. 7, from Simplicius, *Physics* 111.18ff.)

Melissus here spells out some of the consequences of Parmenides' denial of what *is not*. It is not necessary for our present purpose to comment on the whole of the passage: the relevant part is the second half, numbered 7–10. First, Melissus asserts that 'what is empty' is nothing, and hence deserves the description 'what *is not*.' This is presumably an assertion from intuition, or from common speech: if a box is empty, we say there is nothing in it. Of course the box that we call 'empty' is not nothing: it is its contents that we designate as nothing. When the box is empty, we can put something in it; when it is full, we can't. If there is nothing empty anywhere, there is no place for anything to go. Melissus talks about the impossibility of internal change of place, within what *is*. He has already shown to his satisfaction, in fragments that belong to an earlier place in his consecutive argument, that there is only one thing, and therefore has no need to consider whether any external thing could move. But the principle, if established at all, is equally lethal to internal motion and to the motion of something from outside.[10]

[10] In a very interesting paper, 'Two Conceptions of Vacuum,' *Phronesis* 27 (1982), David Sedley has argued that the idea of vacuum as space in which things move is a later invention, in the fifth century. Melissus and the Atomists thought of vacuum, not as space, but as an occupant of space. Some parts of space are occupied by bodies, and some by nothing. But I doubt if this distinction between kinds of vacuum can be maintained. The difficulty is to give an account of the difference between the space and its occupant, when it is not filled by a body. In any case, it does not seem to me to make much difference to Melissus' argument: he speaks of yielding 'into the empty,' and the impossibility of yielding if there is no 'empty.' Sedley points out that the verb ὑποχωρῆσαι (to yield) is not the same as 'to move.' But it is a species of movement, after all. The essential point is Melissus' claim that there must be some place where there is no being, if there is to be the possibility of local change in being.

'The empty,' Melissus argues, is nothing; and nothing is non-existent. Hence there can be no 'yielding' in what *is*, no motion of its parts or within its whole self. We shall see in the next chapter that the Atomists took this very seriously.

The third of Melissus' inventions is again contained in a relatively long piece of connected prose. According to Simplicius, it followed closely on the previous fragment.

(1) That argument, then, is the biggest sign that there *is* only one; but the following are signs too.

(2) If there were many, they must be such as I say the one is. For if there is earth, water, air, fire, iron, gold, living and dead, black and white, and the rest of the things that men say are true – if there are these, and we see and hear them correctly, each must be such as they first seemed to us, and they must not change or become different; each must be always as it is. Now, however, we say that we do see and hear and understand correctly.

(3) But it seems to us that the hot becomes cold, and the cold hot, and the hard soft, and the soft hard, and the living dies and again comes-to-be out of what is not alive, and all these things become different, and what was and what now is are not alike, but the iron, which is hard, is rubbed away by contact with the finger, and so is gold and rock and everything else that seems to be hard, and from water comes earth and rock. [So it follows that we do not see or know the things that are.[11]]

(4) Now these are not consistent with each other. We say there are many, having eternal forms and power,[12] but all seem to us to be altered and to change from what is seen at any time.

(5) It is clear, then, that we did not see correctly and that the former plurality did not appear to us correctly. For they would not change, if they were true, but each would be such as it seemed to us. For there is nothing stronger than what is true.

(6) If it changed, there was death of what is, birth of what is not. So, then, if there were many, they must be such as I say the one is. (Melissus fr. 8, from Simplicius, *De caelo* 558.19–559.12.)

This is a good specimen of the Eleatic challenge to the cosmologists. It depends upon their acceptance, before the argument begins, of the basic premiss that what *is not* is unthinkable and hence to be excluded from discourse, and their acceptance of the first inference from this, that coming-to-be and perishing are likewise unthinkable. The argument then makes the following moves:

A. If we claim that the plural and different things that we perceive (assuming

[11] Barnes, *The Presocratic Philosophers*, vol. 1, p. 340, n. 3, produces a good argument to show that this sentence must be a marginal gloss.

[12] I follow Calogero, *Studien über den Eleatismus*, pp. 89–90, and Reale, *Melisso*, p. 402, in deleting καὶ between ἀίδια and εἴδη.

that seeing and hearing are mentioned as samples) truly exist, then, if there is no coming-to-be or perishing, they must be for ever just as we perceive them now.

B. But we perceive them as changing.

C. Since they do not change, our perceptions in B are false.

D. Hence, we have no reason to believe our perceptions in A either. Perception gives us no reason to believe in the existence of plural and different things.

Melissus expresses the conclusion in language that was to prove significant: 'if there were many, they must be such as *I say* the one is' – that is to say, they must be eternal and admitting no variation in quality. Leucippus and Democritus could claim that their basic elements, the atoms, met these requirements.

9 Leucippus and Democritus

9.1 *The men and their books*

The writers of the ancient world were confident that Leucippus and Democritus were the founders of the atomic theory, but they were not at all careful in allocating the credit between the two of them. Attempts have been made by modern scholars to isolate the contribution of Leucippus – the earlier and less well-known of the two – but their arguments, if not demonstrably wrong, are too tenuous to rely on. Little attempt will be made in this book to distinguish their doctrines, at least with regard to the physical world; on the subject of human society and ethics, it appears that Democritus was very much the major contributor.[1]

The early Atomists are associated with the remote country town of Abdera, on the northern Greek mainland close to the island of Thasos. Democritus lived and worked there; probably Leucippus did so too, although ancient biographers were uncertain whether to connect him with Abdera, Elea, or Miletus. Democritus had a long life – some say he lived to be a hundred. His precise dates are not known. He himself wrote in *The Small World Order*, according to Diogenes Laertius (IX.41), that he was forty years younger than Anaxagoras. Probably he was born about 460 B.C., and Leucippus was somewhat older. The atomic theory, then, must date from about the time of the Peloponnesian War, in the last decades of the fifth century; it is approximately contemporary with Socrates' philosophical activity, and with the birth of Plato.[2]

No works of Leucippus or Democritus have survived. They were engulfed in the surge of disapproval generated by the great Athenian schools. Our knowledge of them depends on quotations in surviving authors, criticisms by their rivals, and summaries by the ancient historians of philosophy. There is a long list of the titles of the books of Democritus, arranged in 'tetralogies' like the dialogues of Plato, in Diogenes Laertius' *Lives of the Philosophers*, IX.45–9: it includes works on ethics, physics, mathematics, music, and applied sciences. Among the

[1] For an extended effort to distinguish the two, see Bailey, *The Greek Atomists and Epicurus*, ch. 2, Leucippus, and ch. 3, Democritus.

[2] Plato is completely silent about Leucippus and Democritus – a puzzling circumstance that will be discussed later.

physical works in this list are '*The Great World Order*, which the Theophrastian school attributes to Leucippus. *The Small World Order*, and *Cosmographia*.'[3] There is also a book called *On Nature*, and several collections of 'Explanations' (*Aitiai*) of features of the physical world – according to one anecdote, he once said that he would rather discover a single explanation than be king of Persia (fr. 118).

But this list remains nothing but a bare catalogue: very little of the information we have about Democritus' argument is assigned by our sources to any particular title. There is therefore no hope of following Democritus' order of exposition; instead, we shall impose our own order, and first discuss what Democritus took from the Eleatics and how he proposed to break out of the box in which their arguments aimed to imprison the physicists, and then consider the more important features of the Atomist cosmology.

The total disappearance of all the many books of Democritus is one of the most lamentable literary catastrophes of the classical world. It is clear that Aristotle had access to whatever he wanted of Democritus' works (he wrote a monograph on Democritus – also lost, of course): presumably he handed over his Democritean library to Theophrastus, and it formed part of the famous Peripatetic collection that went, after Theophrastus' death, to Skepsis in the Troad, from which it eventually made its way to Rome in the first century B.C. Cicero was able to comment on Democritus' style. Sextus Empiricus quotes from one book by name; there are a few other mentions of particular titles in the early centuries A.D.[4] But what is most important is that Simplicius, in his commentaries on Aristotle's *Physics* and *De caelo* – the source of our best information about Parmenides, Anaxagoras, Empedocles, Zeno, and Melissus – makes at the most two direct quotations from Democritus (frs. 167 and 168), and they are very brief and may well come from an intermediate source. It seems that Simplicius, in the sixth century A.D., either had no access to, or took no interest in, Democritus' books.

We have already mentioned, in chapter 1, that the loss of so much of the 'Infinite Universe' literature is no accident. In the era before printing, copies were made very selectively; and the learned world vastly preferred the 'Closed World' cosmology of Plato and Aristotle. This was especially so in the centuries when learning was concentrated among Christians, but

[3] I presume the distinction between 'Great' and 'Small' is an editor's distinction. I doubt whether Democritus adopted the epithet 'Small' himself (out of modesty, Guthrie suggests), and therefore it seems to be unlikely that the title in any way refers to the idea of man as microcosm.

[4] Cicero, *De oratore* I. 11.49; *Orator* XX.67; Sextus, *Math.* VII.136–8; Proclus, *In rempublicam* II.113.6.

in Simplicius' neglect or ignorance of Democritus' writings we can see clear evidence that the judgement against Atomism was made independently of Christianity.

9.2 *What is and what is not*

Like Anaxagoras and Empedocles, the Atomists accepted the proposition that nothing that *is* comes into being out of what *is not* or passes away into what *is not*. We have seen in earlier chapters that this proposition is not to be interpreted only in the innocuous sense that nothing comes into being out of *nothing*, but also in the strong sense that nothing comes to be out of what is not *it*. Anaxagoras had tried to meet this requirement with his theory that everything that apparently comes to be was previously latent in something else; Empedocles tried to meet it with the assertion that everything that *is* can be reduced to earth, water, air, and fire, which themselves 'are for ever changeless in the cycle.' The Atomists then followed Empedocles in claiming that everything that *is* is reducible to changeless elements, but they differed from him about the nature of the elements. We must try to explain what the differences were, and why the Atomists thought they were required.

The most important difference is that the Atomists banished the traditional 'opposites' from the elements altogether. Hot, cold, dry, wet, and other perceptible qualities are relegated to a lower rung on the ontological ladder. In Empedocles' system, the element fire is hot and dry; the element water is cold and wet. But according to Democritus no *element* can be said to be hot or cold or dry or wet. The elements are indeed plural, and we shall see how they are to be distinguished from each other; but all are of stuff of exactly the same quality, or lack of quality. He expressed this idea in paradoxical form in a saying quoted by Sextus Empiricus, and again by Galen (fr. 9, fr. 125):

> By custom, sweet; by custom, bitter.
> By custom, hot; by custom, cold.
> By custom, color.
> In truth – atoms and void.

Democritus uses the same expression, 'custom' (*nomos*), that we found Empedocles using of 'coming-to-be' and 'death' (above, p. 84), and he uses it in just the same sense. When there is a coming together of previously separated elements, said Empedocles, people speak of 'coming-to-be,' and when they separate again, they speak of 'death' – and this, he said, was a *custom* that he proposed to follow. Similarly Democritus proposes to show that the perceptible qualities are reducible to the elements, atoms and void. The customary names do not pick out permanent and irreducible beings, but only objects that must be analyzed into properties of atoms and void.

Why was this change made? Perhaps it was the argument of Melissus that necessitated it. As we have seen in chapter 8, the point of his argument in fragment 8 was this: if we distinguish earth, water, air, fire, iron, gold, living, dead, black, white, and so on, and believe these to be *real* distinctions, the reason why we do so is that we trust our senses. But the same senses report that the hot *becomes* cold, the hard *becomes* soft, the living *becomes* dead, and so on. If the initial report of the distinctions was right, then the second report of the changes should be trustworthy too. But on Eleatic principles the latter cannot be right: as Melissus puts it, 'if there is a change, then what *is* has perished and what *is not* has come to be' (fr. 8.6). Hence, the first report, too, must be untrustworthy. If the Atomists read and accepted this argument, it would constitute a reason for stripping away from the elements all the *perceptible* qualities, if we grant that all perceptible qualities are liable to change.

Now, in the earlier pluralist systems, differences in quality served as the marks by which one part of the physical world could be distinguished from another. In Anaxagoras' cosmos, initially no internal distinctions were possible, because 'all things were together'; but after Mind's revolution had sorted out the ingredients somewhat, distinctions could be made according to the qualities that predominated here or there. In Empedocles' world, similarly, when the elements had been sorted out from the total mixture of all things in the Sphairos, they enabled one region to be distinguished from another by their qualities, according to their proportions in each region. But in the Atomist system, since the elements lacked qualities, a new kind of differentiation had to be found: without that, their universe would be like the One continuous Being of Parmenides' Way of Truth – a single, homogeneous being within which no change can be detected because no part can be distinguished from the rest.

To distinguish one part of what *is* from another – in other words, to account for plurality – Leucippus and Democritus introduced the concept of void. Each unit of being – each atom – had a shape and a boundary; between its boundary and the boundaries of other atoms was an interval in which there was nothing at all. Since the concept of the void plays a most important role in the controversies of physicists from the time of Leucippus onwards, we must go back on our tracks to find out its origins and the course of its development.

The intuitive notion of something *empty* is, of course, older than the earliest literature. In the *Iliad* (III.376) Menelaus grabs hold of the helmet of Paris and drags him away, until the chinstrap breaks, and he tosses away the *empty* helmet in disgust. 'Empty' means 'not containing the *relevant* filling.' As we saw in chapter 5, some early Pythagoreans were said by Aristotle to have put up a theory according to which the world grew when a primitive One turned itself into a plurality 'by breathing in the void' – but the very expression 'breathing in' shows that they had not yet distinguished between a space empty of everything but air, and a true void. That distinction was made clearly in the time of Anaxagoras and Empedocles: both wrote of 'experiments' with homely devices that showed that what one might call an 'empty' vessel was not truly empty. 'They show that air is something,' says Aristotle (*Physics* IV.6, 213a25–27), 'by distending wineskins and showing that the air has strength, and by catching it in clepsydras.' The 'empty' wineskin, inflated and stoppered, will bear a man's weight; the clepsydra (a hollow vessel with a strainer at the bottom and a narrow vent at the top, used for transferring liquids from one container to another without moving the containers) will not let any liquid in when you immerse it, 'empty,' into the jar, if the top vent is plugged, because it is full of air.

The first reaction to the distinction between a true void and a vessel filled with air was to deny that there is ever a true void, anywhere in the world. This was the position of Parmenides and his Eleatic followers. A true void is 'nothing,' and to talk of nothing is to talk to no purpose. As Melissus wrote: 'And nothing is empty; for what is empty is nothing. Now, what is nothing could not *be*' (fr. 7.7). Anaxagoras and Empedocles both agreed with the Eleatics about this. So although it seems clear that the concept of a true void was understood before the Atomists of Abdera built it into their system, it is likely that Leucippus was the first to assert the existence of a void.

One might have expected that the Atomists would rebut Melissus by showing that the void is not to be identified with 'nothing.' Instead they took the extraordinary and paradoxical way of accepting that identification, and then making a distinction within the concept of 'nothing.' Plutarch in his *Reply to Colotes* (1109A) quotes some words of Democritus 'in which he declares that the *-hing* (τὸ δέν) *is* no more than the *not-hing* (μηδέν) – giving the name "*-hing*" to body, and "*not-hing*" to the void – meaning that the void too has its own existence and nature.' What exactly Democritus was up to with this strange wordplay is a matter of guesswork, but it would be appropriate if he had something like this in mind. We are accustomed to thinking of 'nothing' as *negating* a sentence: 'I hear nothing' means 'I am not hearing anything,' 'no hearing is being accomplished by me.' By suggesting that the word is to be split in this odd way, which we can render in English by writing 'not-hing' instead of 'no-thing,' perhaps he meant to indicate that 'nothing' is not always a negation, but sometimes a negative predicate, of the form 'not-x.' Thus he would make the point that the void is not *non-existent*, but rather is *not anything in particular*. If we read 'nothing' as a negation, then the Atomists' proposition 'A is separated from B by nothing' is equivalent to 'A is not separated from B'; but if we interpret it as a negative predicate, the sentence means 'A is separated from B by (something that is) not-anything-in-particular.'[5]

But one may well ask why Leucippus should take this elaborate way of introducing the void; why not give up the expression 'what *is not*' and 'nothing' as designations of the void altogether, and just use 'the void,' as the Epicureans were to do? Again, one can only guess at the answer to this, since there is no explanation in the fragments. Probably the answer is still to be found in the Eleatic connection. It may have been considered that to posit two kinds of *beings*, the full and the empty, would be to reintroduce the perceptible opposites that had been ruled out by Melissus' argument, discussed above in chapter 8. Another possible reason can be found in an argument of Zeno (fr. 3). Zeno argued that any two items of being need a third item to separate them, and that third needs another item to separate it from the first, and so on *ad infinitum*. But if the third item, that separates the first two, is *not-being*, then no regress is generated by the proposition that any two items *of being* need something to separate them.[6]

Void was needed in the Atomists' system, then, to separate one piece of being from another. It was also needed to allow motion to take place. Melissus had argued thus:

> And nothing is empty. For what is empty is nothing. Now, what is nothing could not *be*. Nor does it move. For it cannot yield anywhere: it is full. If it were empty, it would yield into the empty; there being nothing empty, it has no place to yield. (fr. 7, §7)

Leucippus and Democritus denied that the void, being nothing, must be non-existent; they agreed that *if* there were no void there could be no motion, but since they denied the antecedent clause, they were able to reject the Eleatic's conclusion and allow that there is real motion of real things.

[5] For further discussion of this point about μηδέν, see Moorhouse, *CQ* 15 (1965) and Mourelatos, '"Nothing" as "Not-Being,"' in Bowersock *et al.*, *Arkturos*.

[6] This point is due to Stokes, *One and Many in Presocratic Philosophy*, pp. 221–2.

Epicurus was to make use of Melissus' argument, too:

That there are bodies, is attested in all cases by perception itself, through which we must infer what is unclear, by means of reasoning, as I said before. If there were not that which we call *void* and *space* and *intangible nature*, bodies would have nowhere to be and nowhere to move, as they are seen to move. (*Letter to Herodotus*, 40)

It is important to notice that Epicurus plainly takes as his premisses in this argument propositions drawn from sense-perception: bodies exist and bodies move. From these, together with Melissus' argument, he deduces that there is void. This is consistent with his epistemology, which defends the reliability of sense perception. But Leucippus and Democritus adopted a more cautious position about perception (see below, pp. 131–5): the evidence is confused, but it is clear at least that we should be more likely to find an *a priori* argument for the existence of void than an empirical one. And indeed the primary evidence for their argument does not give sufficient grounds for attributing the Epicurean argument from motion to Leucippus and Democritus. The evidence is a passage of Aristotle:

But Leucippus thought he had arguments that would assert what is consistent with sense perception and not do away with coming-into-being and perishing and the plurality of existents. He agrees with sensible appearances to this extent, but he concedes to those who maintain the One [*sc.* the Eleatics] that there would be no motion without void, and says that the void is what *is not*, and that no part of what *is* is what *is not* – for what *is* in the strict sense is wholly and fully a being. But such a being, he says, is not one; there is an infinite number of them, and they are invisible because of the smallness of their mass. They move in the void (for there *is* void), and when they come together they cause coming-to-be, and when they separate they cause perishing. (*De generatione et corruptione* 1.8, 325a23–32)

What lies behind this seems to be not the Epicurean argument 'no motion without void; but we observe motion; therefore there is void,' but rather a defensive argument against the Eleatics, of this form: 'no motion without void; but there is void; therefore there can be motion.'[7]

Before we turn back to the other aspect of the Atomists' universe, from what *is not* to what *is*, or body, there is one more question about the void

[7] This corrects what I wrote in 'Aristotle and the Atomists on Motion in a Void,' in Machamer and Turnbull, *Motion and Time, Space and Matter* (to be reprinted in my *Cosmic Problems*), where I attributed the Epicurean argument to Leucippus and Democritus.

Sedley, *Phronesis* 27 (1982), has argued that we should distinguish between void conceived as an occupant of space, and void conceived as the space that is occupied, or not, by body. He would attribute the first to Leucippus and Democritus, and the second to Epicurus. But this means rejecting Aristotle's evidence (*ibid.* p. 179), and I am not convinced that we are justified in doing so. See also chapter 8, note 10, above.

that needs an answer. If the Atomists had found a way to admit the existence of what *is not*, why did they need to admit the force of *any* of Parmenides' arguments in the Way of Truth, or the later arguments based on them? The basic premiss of those arguments was just this: 'you cannot know what *is not* (that is not to be accomplished) / nor utter it' (Parmenides, fr. 2.7–8). This premiss was used in the argument that what *is* cannot come into being or pass away, and again in the argument that what *is* is a continuous whole that admits no difference of quality or of degree. If the premiss is denied, what remains of the arguments? Why were the Atomists still concerned, like Anaxagoras and Empedocles, to deny coming-to-be and change at the fundamental level?

I think the answer must be that the Atomists did not exactly deny the premiss, but only modified it, in such a way that the dependent arguments still appeared to hold. They would have been compelled to deny some of Parmenides' expressions of the premiss, such as 'Never shall this prevail, that things that *are not, are*' (fr. 7.1), since that is precisely what did 'prevail' in their conception of the void. But some force would still remain in 'you cannot know what *is not*.' What *is not* – the void – is still only the negation of whatever properties belong to what *is*: it is without properties of its own. This is particularly significant in connection with Parmenides' argument against the coming into being and passing away of what *is*. How could what *is* lose its properties? How could they simply be annihilated? Correspondingly, how could they arise from an absence or a privation? That is what remains unknowable and unutterable.

On this analysis it becomes clear that although the Atomists admitted the being of what *is not*, they had not yet progressed as far as Plato in the *Sophist*. They had not reached the notion of 'is not' as equal to 'is other than,' and so they were not yet ready to admit that a subject that *is* may nevertheless bear innumerable predicates containing 'is not.' As Aristotle put their position, 'no part of what *is* is what *is not* – for what *is* in the strict sense is wholly and fully a being' (*De generatione et corruptione*, 1.8, 325a28, quoted on the last page). They could not agree with Plato (*Sophist* 256e): 'Concerning each of the Forms, there is much of what *is*, but an unlimited multitude of what *is not*.'

9.3 *Atoms*

To anyone looking back from the twentieth century to ancient Abdera, the origins of the concept 'atom' must appear paradoxical almost beyond belief. No experiments contributed to it, and remarkably little in the way of observation. No argument in the remotest degree like those of twentieth-century physics or chemistry was used in setting it up. Nothing in its first appearance could give a hint of its future role in the politics of the human species. Yet an unbroken chain links Democritus of Abdera and Rutherford of Cambridge: the chain runs from Democritus to Epicurus and Lucretius, and from the Epicureans to Gassendi; from Gassendi to Charleton, Boyle, and Newton; and so to Dalton and the vast edifice of atomic theory built in the last century and a half. Newton, standing at the beginning of modern physics and yet somehow still close to classical antiquity, makes the continuity more intelligible:

> It seems probable to me that God in the beginning formed matter in solid, massy, hard, impenetrable, moveable particles, of such sizes and figures and with such other properties and in such proportion to space, as most conduced to the end for which he formed them; and that these primitive particles being solids are incomparably harder than any porous bodies compounded of them, even so very hard as never to wear or break in pieces. (*Optics*, ed. Horsley, IV.260)

If the creator God and his teleological activity were removed, Democritus could hear an undistorted echo of his words in this passage.

The Greek word *atomos* is an adjective, meaning 'uncut' or 'indivisible.' I believe the only instance of the adjective in literature earlier than Democritus is in Sophocles' *Trachiniae* 200, where it is used of grass and means 'unmown.' Curiously enough, it occurs only once in passages that are thought to be in the original words of either Leucippus or Democritus; there it functions as a neuter noun, in the plural: 'indivisibles' (Democritus fr. 9).[8] It is attributed to them both frequently in secondary sources, and it came to be the most usual word in Greek for indivisible material elements.[9]

We may guess, although no ancient source tells us this in so many words, that the primary reason for asserting that there are indivisible bodies is drawn from the argument against physical change in Parmenides' *Way of Truth*. The line of thought is this. If *two* come into being where *one* was before, then plainly some new 'thing that *is*' has appeared on the scene. But how could this be? We have not quoted this part of Parmenides' poem before; so it will be as well to have it here:

> It *is* now, all together
> one, continuous. For what origin of it will you seek?
> How and whence grown? Not from what *is not* – I will not grant
> that you could say or know that, for it is not to be said or known
> that it *is not*. And what need could have stirred it

[8] Fr. 9 is from Sextus; the same words are repeated in fr. 117 (Diogenes) and fr. 125 (Galen).

[9] As well as the neuter, we find the feminine. The neuter suggests the noun σώματα, the feminine suggests φύσεις or possibly ἰδέαι (Plutarch, *adv. Coloten* 1110F).

> later or sooner, starting from nothing, to grow?
> Thus, either it must *wholly be*, or not.
> Nor will strength of conviction allow that something comes to be
> beside it, from what *is not*[10]. (Parmenides fr. 8.5–13)

We may suppose, then, that the Atomists interpreted the division of a single piece of what *is* into two as involving the coming into being of some new being (the second), or else perhaps as the coming into being of two new beings along with the destruction of the single being. In either case, it was to be rejected: there was to be no *transition* from being to not-being, or vice versa.

> They claimed that a plurality could not come into being from what is truly one, nor a one from what is truly many: that was impossible.[11] (Aristotle, *De generatione et corruptione* 1.8, 325a34)

This is a line of argument to the effect that there *must* be indivisible bodies, if there are any unitary bodies at all. It is still necessary to locate them, so to speak, in the physical world. At first sight, it might appear that we have empirical evidence that *all* bodies are divisible: even the hardest things, as it seems, are worn away in time – rocks are hollowed out by dripping water, rings are worn thin by rubbing against the skin of the finger. These examples had been cited by Melissus as indicating the apparent perishability of material objects (fr. 8). For this reason, if there are to be indivisibles, they must be located below the level of perception, and a physical property must be given to them to explain how it is that they can be indivisible, when perceptible bodies manifestly are not. This was found simply in the solidity of the atoms. To put it crudely, Democritus relied on the notion that only what is cracked can be broken: if a thing is divisible, it must be because there is void in it. So the Atomists postulated the existence of bodies, too small to be perceived, in which there is no void at all, and they were able to claim that the perfect solidity of these bodies made them indivisible.

But there was another reason for postulating indivisibles. Aristotle gives at length what he describes as 'the argument that appears to prove that there are indivisible magnitudes,' in the second chapter of *De generatione et corruptione*.[12] Although he does not say in so many words that

[10] If the last line and a half are to say anything that is not said in the previous lines of this quotation, then either ἔκ γε μὴ ὄντος in 12 must be emended to ἔκ γε τοῦ ὄντος or ἔκ γε δὴ ὄντος (so Reinhardt, Tarán, Hölscher, and others), or at least παρ' αὐτό must be taken to mean 'beside what *is*' (not 'beside what *is not*'), as my translation tries to have it.

[11] For a good discussion of the cluster of passages on this theme in the Atomists, see Stokes, *One and Many*, pp. 225–36.

[12] I have discussed this argument in *Two Studies*, pp. 83–5, reprinted in Mourelatos, *The Pre-Socratics*, pp. 508–11. I shall not do more than summarize that discussion here.

the argument was framed by Democritus, and indeed inserts some of his own terminology into it, he begins it by congratulating Democritus for 'having been persuaded by proper, physical arguments' – as opposed to the Platonists, who used abstract arguments to attain a similar conclusion. So in spite of the doubts of some scholars, I think we can regard the argument as being genuinely Democritean.

The argument is an indirect one. It begins by supposing the contradictory of the assertion that atomic bodies exist – that is, that body is divisible *everywhere*, or *through and through*. Then we ask what happens if such a body is actually divided through and through. Three possibilities seem to exhaust the range: (1) that the product of the division is a collection of units each having some magnitude; (2) that we have a collection of units each having no magnitude; and (3) that we have nothing left at all. But each appears to be impossible: (1), because it contradicts the hypothesis that the body was divided through and through, (2) and (3), because the body must be composed of the units into which it is divided, and it could not be composed either of units which have no magnitude or of nothing at all. Hence, the argument concludes, there must be indivisible bodies. Aristotle goes on to show that the argument does not work, because there is another way out of the difficulty: there is a way in which a body may be *divisible* everywhere, without ever being *divided* everywhere so as to yield these problematical units.

The way in which the argument is presented in the *De generatione et corruptione* has given rise to a controversy about how to interpret Democritus' atomism. Aristotle's subject in this book is what happens to physical body, or 'sensible body' as he calls it (1.2, 316b18). If the argument is taken to apply only to the physical level, then it would appear at first sight that the indivisibles in question must be unsplittable lumps of matter – indivisible just in the sense that it can never be the case that two adjacent parts come to be separated from each other by an interval. On this interpretation, what we have here is another argument for the existence of material atoms, and it is not entailed that they have no parts whatever, whether separable or inseparable. But this interpretation has some very unsatisfactory features about it. The case against it rests fundamentally on two propositions: first, the argument reported by Aristotle is evidently designed to meet Eleatic arguments about divisibility, as we shall see very shortly, and those arguments would not be rebutted merely by postulating physically unsplittable bodies; and second, Aristotle's own arguments against the Atomists, in this same chapter of *De generatione et corruptione* and in book VI of the *Physics*, are directed against 'indivisible *magnitudes*' (i.e. partless units of extension) rather than unsplittable bodies.

These objections to a purely physical interpretation suggest that the Democritean atom was not merely a body that could never be split into two or more parts, but actually one that had no parts at all. That means that to talk about the parts of an atom is to contradict oneself: it is like talking about even triplets or married bachelors. (It is a position that can be distinguished from 'geometrical atomism,' as we shall see in section 9.4.) But there are objections to this interpretation, too, which many find even more powerful than those on the other side. The chief difficulty lies in the fact that it seems impossible to give a rational account of differences of shape in partless atoms, and yet atoms of different shapes were crucial to the theory. We shall return to this very difficult problem in the next section.[13]

It is plain that the argument reported by Aristotle is constructed out of materials lying to hand in Zeno's arguments against pluralism, discussed in chapter 8. We can arrange Zeno's arguments in the following frame:

1. If what *is* can be divided everywhere, so as to produce ultimate units having no size, then it has no size itself (fr. 2).
2. But if what *is* can be divided everywhere, so as to produce ultimate units having size, then, since the units are infinitely numerous, it has infinite size (fr. 1).
3. But if what *is* can be divided everywhere, without ever reaching ultimate units, such a being can never be traversed (the Dichotomy and Achilles arguments).

The conclusion that Zeno proposed to draw from this was that what *is* cannot be divided, i.e. that there is no plurality of things that *are*. The Atomists countered by pointing out, in effect, that the premiss that is refuted by Zeno's *modus tollens* arguments is not 'what *is* is divisible,' but 'what *is* is divisible *everywhere*.' If we substitute this former premiss in the arguments just outlined, none of Zeno's conclusions hold. Thus in (1), if what *is* is divisible into partless atoms, each having finite size, it does not follow that the original 'what *is*' has no size itself. In (2), if we start with a finite collection of atoms as our 'what *is*,' division into the component atoms will produce only a finite number, and hence it does not follow that the original collection was infinite; if we start with an infinite collection of atoms, that conclusion does follow, but the Atomists would admit in any case that such a collection must be infinite in size. In (3), by postulating the existence of partless atoms the Atomists have denied the premiss of the infinite divisibility: they have denied the possibility of an *infinite* sequence

[13] There is a long and continuing controversy about this; for more details see ahead, notes 17–19.

of steps to be taken before reaching a goal. The man crossing the stadium, and Achilles pursuing the tortoise, will be separated from their goal by smaller and smaller stretches of territory until they are separated only by one partless atom: the next move closes the gap, and they have reached their goal.

This strategy on the part of the Atomists does not constitute a *solution* of Zeno's puzzles about divisibility. Instead of solving them, it claims that they can be ignored, because the situation envisaged in the puzzles does not obtain in the real world. What *is* is simply not divisible in such a way as to give rise to these puzzles. Aristotle was correct in saying that they 'gave in' to the Eleatic arguments by positing 'atomic magnitudes' (*Physics* I.3, 187a1). They gave in, in the sense that they accepted the force of Zeno's arguments: *if* what *is* is divisible everywhere, then these paradoxical conclusions follow. Accepting the arguments, they asserted that what *is* is not divisible everywhere: atoms are parts of what *is*, and they are not divisible.

9.4 *Atoms and mathematics*

We have seen that there were many properties denied to the atoms of Democritean theory out of respect for Eleatic arguments: thus atoms were not destructible, not changeable, not divisible in any sense, and not characterized by heat or cold or color or other perceptible qualities. What properties, then, could be attributed to them without infringing the Eleatic rules that the Atomists were committed to?

The atoms have body, or resistance, as we have observed, but without any difference in the *kind* of body. They were held to differ from each other, however, according to the *quantity* of body that they had. That is to say, they were of different sizes and shapes. The range of these differences is a controversial matter. Some evidence suggests that Democritus postulated some very large atoms, but I prefer the evidence of Aristotle, which says that all atoms were supposed to be imperceptibly small.[14] Differences in size and shape among the atoms, however, were a vital ingredient in the explanation of perceptible qualities; so we have to ask whether it was possible for Democritus to give a coherent account of these differences,

[14] Passages suggesting very large atoms are Aetius I.3.18, in DK 68A47 (the single word κοσμιαῖαν, 'like a cosmos'), and Eusebius, *Praep. Evang.* XIV.23.2.3 in DK 68A43 (as opposed to Epicurus, Democritus said some atoms would be very large). On the other hand Aristotle says, in *De generatione et corruptione* I.8, 325a29, and in fr. I (Ross) of his lost book *On Democritus*, that Democritean atoms were 'so small as to escape our senses.' In *Two Studies*, p. 96, I suggested a reason why the errors might have arisen. The argument to this effect is developed at length by O'Brien, *Theories of Weight in the Ancient World*, vol. I, pp. 282–98.

while continuing to respect the Eleatic arguments. Was this something the Atomists could do consistently with the rest of their theory?

We must return to a subject introduced in the last section: the difficulty of reconciling the different shapes of the atoms with their partlessness. Suppose we have two atoms, one shaped like an F and the other like an E. Can we account for the difference without referring to a part of the second that is lacking in the first? That this was thought to be a difficulty in the theory is shown by the modification introduced by Epicurus, who dissociated the physically indivisible from the theoretically indivisible, and asserted that the atom contains indivisible parts. But how Leucippus and Democritus regarded this problem it is impossible to decide. Possibly they would have fallen back on the weak argument that we are just wrong to talk about a part of the E that is lacking in the F, since the F never had such a part and the E can never lose it. To talk of this part is to isolate it – but it has no existence in isolation. Such talk, they might have said, about isolated parts of atoms, like the geometricians' talk of a line touching a circle at a dimensionless point, corresponds to no possible fact in the world.[15] Hence, if it leads to paradoxes such as Zeno's, we should be neither surprised nor disconcerted.

The problem remains, however; how can we *avoid* talking about parts of atoms, if we want to describe an atom shaped like an E, or shaped in any way whatever? What is it to have a shape, except to have parts distributed in this manner or that? To this objection, again, it is possible to construct an answer of sorts, with only a mite of evidence that the Atomists adopted it. If we make a distinction between parts that, when put together, add up to the whole (quantum parts), and the extremities or limits contained in or containing the whole, then we might give an account of shape without mentioning parts. We are told by Aristotle that Democritus described the sphere as a kind of 'angle'; Simplicius comments: 'If what is curved is an angle, and the sphere is curved all over the whole of itself, then it is reasonably called a whole angle' (DK 68B155a). Could this be a fragment of the enterprise of defining atomic shapes entirely by outline, to avoid discussing the arrangement of parts?[16]

[15] Aristotle, *Metaph.* III.2, 997b35: 'Neither are perceptible lines such lines as the geometer speaks of (for no perceptible thing is straight or curved in this way); for a hoop touches a straight edge not at a point, but as Protagoras said it did, in his refutation of the geometers' (trans. Ross). Of course, Protagoras was contrasting the perceptible world with the geometer's; an extra step is involved if we contrast the imperceptible atom with geometrical figures.

The list of Democritus' writings in Diogenes Laertius IX.47 contains the title *On difference of cognition or on the contact of circle and sphere*. This suggests that he tackled the same question as Protagoras, but it gives no clue as to his attitude to it.

[16] This is to take the point of fr. 155a in a totally different sense from (e.g.) Guthrie (see

To get this subject clearer, we must return to the distinction between different kinds of divisibility (see above, pp. 125–6ff.).[17] It is clear that Democritus' atoms were physically indivisible, in the sense that no atom could be composed of, or broken into, smaller bits. The postulation of atoms in this sense differentiated Democritus from Anaxagoras and, I think, from Empedocles: both of these adopted accounts of the nature of matter that managed without indivisible corpuscles. Not all scholars share my belief, however, that Democritean atoms were also theoretically indivisible, in the sense that they contained no parts at all, not even inseparable ones.[18] This position divided Democritean from Epicurean atoms. But perhaps even if one grants that Democritean atoms were partless, there remains a further question to be discussed: whether Democritus was also a *geometrical atomist.*

It would have been possible, in theory, for Democritus to accept the existence of conventional geometry as the science of continuous and divisible magnitudes, and to deny that this science applied to atoms: perhaps it applied only to the void, or perhaps it was a construction with no application to the physical world at all. Secondly, it might perhaps be the case after all that he could find a way to preserve both the physical and the theoretical indivisibility of atoms while at the same time holding that a conventional geometrical account could be given of their shape.[19] Thirdly he could dismiss conventional geometry as based on a false premiss, and construct an alternative atomistic geometry.[20]

Against the second of these possibilities, I think it is almost decisive that Aristotle believed Democritean atomism to be in conflict with mathematics (*De caelo* III.4, 303a20–4). The claim is sometimes made that it was Aristotle's own theory of the nature of geometrical magnitudes (namely

---

HGP, vol. II, p. 485), who thinks it means that what appears smoothly spherical to the senses is actually a polyhedron with invisibly small facets. I do not understand how this could be right, since Aristotle is plainly talking about the shape of atoms.

[17] See also Sorabji, *Time, Creation, and the Continuum*, pp. 350–7. There is a useful discussion of the question, with a handy collection of the evidence, in Löbl, *Demokrits Atome*, pp. 226–43.

[18] This used to be denied on the ground that it entailed an atomistic geometry, and Democritus 'was too good a mathematician' to believe in that (Sir Thomas Heath, *The Thirteen Books of Euclid's Elements*, vol. I, p. 181). It was asserted by Luria, *Die Infinitesimaltheorie der antiken Atomisten*. I followed Luria's line, with variations, in *Two Studies*, pp. 79–103; so did Guthrie, *HGP*, vol. II, pp. 503–7. Jonathan Barnes returns to the denial, in an excellent discussion, *The Presocratic Philosophers*, vol. II, pp. 50–8. But he seems to posit a Democritus who is at the same time sharp enough to distinguish physical indivisibility from partlessness, and confused enough to posit physical indivisibles as an answer to Zeno ('reflecting in a vaguely Zenonian fashion on physical division': p. 58).

[19] This appears to be the position of Seide, *Hermes* 109 (1981).

[20] Something like this is attributed to him by Popper, *Conjectures and Refutations*, pp. 80–4 (in the Harper Torchback ed. of 1968).

that they are abstracted from physical magnitudes) that led him to deduce mathematical consequences from Democritus' physical theory; and hence that his evidence is not to be trusted when reconstructing Democritus' thought. But we should beware of attributing greater clarity about the distinction to Democritus than to Aristotle. It is highly unlikely that Aristotle would have so bluntly identified a conflict here if Democritus had already given reasons for claiming that there was no conflict. And in any case, Aristotle was surely right: if the theorem that the volume of a sphere is divided into two halves by its diameter is applicable to a spherical atom, then it cannot be denied that the atom theoretically contains parts.

Against the third possibility, that Democritus was a geometrical atomist, what counts most is that there is little trace of the postulated alternative geometry. There is a famous and tantalizing paragraph in Plutarch that is often quoted in this connection.[21] Democritus is said to have raised the problem of what happens when we make a horizontal cut through a cone parallel to the base: are we to think the surfaces of the two segments are equal or unequal? If they are unequal, the cone's sides must be stepped, in notches; if they are equal, the cone will have the properties of a cylinder. But if this is an argument for an atomistic geometry, then the puzzle was merely a temporary one for Democritus: the surfaces are in fact unequal, the cone is stepped, and the problem is solved. Plutarch, however, says only that Democritus raised the problem: he does not attribute any solution of it to him. And there is no solid evidence elsewhere for an atomistic Democritean geometry.

The cone argument seems, in fact, to support the first option rather than the third. Democritus raised the cone as a problem. But there is no problem if the cone is made of atoms, as we have seen. On the other hand there is no problem for the conventional geometrician, either: for him, the surfaces of the two segments are plainly equal, since they coincide. The problem arises only if one supposes geometry is applicable to a physical solid.

The balance of the evidence, then, seems to me to show that if Democritus faced the problem raised by the partlessness of atoms and the continuity of geometrical magnitudes, his response was not to abandon either, but to separate their domains. Geometry is not a description of physical being or beings. Is it then a description of the void? If that were the case, we would have to abandon the notion of Democritean void as a space that may or may not be occupied by an atom or atoms: if the atom coincides with a portion of space, and that space is also a geometrical magnitude, then geometry must apply to the atom. So we would be thrown back on

[21] *De comm. not.* XXXIX, 1079E–F.

the other interpretation of void mentioned already in note 7: void would be the interval between the atoms – 'the empty' as opposed to 'the full,' as Democritus expressed the distinction. This may be right, but I am inclined to think the answer is rather that he thought geometry was not applicable to the magnitudes of the cosmos at all, whether full or empty.

9.5 *Perceptible qualities of compounds*

The evidence on this subject is confused and difficult – most unfortunately, since it is a crucial part of the ancient Atomists' system – and to avoid becoming embroiled in philological detail we shall have to be content with summary statements and some sampling.[22] It seems clear that the contents of our perceptions were to be explained by four types of factor: (1) the size and shape of the atoms composing the perceived object, (2) the position, order, and distribution of these atoms, including their distance from each other, (3) the distance and conditions between the perceived object and the perceiver, and (4) the state of mind of the perceiver. We will discuss each of them briefly; but first it is necessary to say something about the mechanics of perception.

Democritus' theory plainly included a mechanical explanation of the interaction between a perceiver and the external world: our difficulty is that we hear of two such explanations.[23] The structure that underlies the whole theory, in all its versions, is a contact between the soul of the perceiver and the atoms of the perceived object, with their various shapes and sizes, or if not these atoms themselves, then some intermediary product of their motion. The simplest case is taste: one taste differs from another because of the different shapes and arrangements of the atoms of food taken into the mouth. The most complex case is vision. Even here there is some kind of material contact, mediated or not, between the atoms of the perceived object and the soul of the perceiver. It is not the object itself, of course, that hits the eye, but at most some kind of effluent from the surface of the object. The effluent is a series of 'images': the Greek word is *eidola* ('idols'), which in earlier Greek literature normally refers to dream visions or ghosts or the 'shades' of the dead (Democritus is also said to have used the rare word *deikela*, 'representations'). The idea is developed fully by the Epicureans, especially in book IV of Lucretius' poem, but the sources give only sketchy information about Democritus' use of it. Evidently, the *eidola* are made of atoms, but are so flimsy that shedding them

[22] The most important source is the long fragment of Theophrastus, *De sensibus* (49–82 = DK 68A135).

[23] The first explanation is given in Alexander, *De sensu* 24.14ff. = DK 67A29. The second is in Theophrastus, *De sensibus* 50 = DK 68A135.

does not noticeably deplete the bulk of the parent object. In Epicurean theory, they themselves are said to impinge on the eyes; the same is sometimes said of Democritus, but we also find a complex account of an impression made upon the intervening air by the effluent of the perceived object – with a contribution made by an effluent from the observer, too. It is this airy impression that enters the eye. We shall have to pass over the difficulties of detail here, and be content with the bare recognition that Democritus' theory did contain an account of this difficult piece of physics.[24]

Different atomic shapes cause different sensations. Thus a sharp taste is said to be produced by atoms that are small and angular, a sweet taste by atoms that are round and not so small. A light color is associated with atoms of a smooth shape, a dark color with rough sharp-angled atoms.

Position and arrangement of the atoms forming a compound are significant in an obvious way. Compounds are made of heterogeneous atoms; but most sense-impressions are of surfaces, and they depend on the nature of the atoms at the surface, which may differ from time to time. The difference between a light and a dark color depends in part on the relative position of the atoms at the surface, light colors being associated with a level surface and dark with unevenness. The quantity of void within the volume of the compound affects tactile qualities such as hardness and softness. In some cases it makes the difference between perceptibility or not; air, for instance, is a rare compound, normally imperceptible, but it may become perceptible if the atoms are packed more tightly together (mist or fog).

Since perception at a distance is brought about by the passage of physical objects – *eidola* or air-impressions – through the intervening space, the sensation is inevitably affected by whatever happens to these flimsy objects in transit. Apart from obvious effects – solid barriers prevent perception altogether, mist makes the image cloudy, and so on – passage through the air is invoked in some way (surviving texts are not clear about the detail) to explain the reduction in size of the visible object to fit the eye of the observer. A possible interpretation of the evidence is that vision requires a contribution from the seeing eye, in the form of a cone-shaped radiation of atoms that move from the eye through the air towards the

[24] For further discussion, see von Fritz, 'Democritus' Theory of Vision,' in *Science, Medicine and History*, vol. 1 (1953); German version in his *Grundprobleme der Geschichte der antiken Wissenschaft* (1971); Bicknell, *Eranos* 66 (1968); Burkert, *Illinois Classical Studies* 2 (1977). Burkert's is an especially penetrating article, although I am not convinced by his conclusion that the air-imprint theory is incompatible with the *eidola* theory and that Democritus simply changed his mind. I believe the air-imprint theory may be just that part of the *eidola* theory that accounts for the reduction in size of the image between object and observer.

eidola coming from the objects in the field of vision. When the two encounter each other, they condense the air between them in such a way that an image is formed in the air. And this image is somehow transported back to the eye along the 'cone' of vision.[25]

The final factor in the make-up of our perceptions is the condition of our psyche itself. Like his contemporary and fellow citizen of Abdera, the sophist Protagoras, Democritus was impressed by the observation that one and the same object could appear differently to different subjects: what tastes sweet to the healthy man may taste bitter to the sick man. If the three factors we have already discussed remain constant, the factor that varies must be in the subject himself. Democritus stands at the beginning of a discussion of this topic that lasts for several centuries. He apparently held the three following theses: that we can be confident of our knowledge of the atomic theory, that knowledge of the imperceptible elements comes indirectly from the senses, and that two perceivers may perceive the same object differently. This rather sophisticated position is summed up in four graphic quotations:

> By custom, sweet; by custom, bitter.
> By custom, hot; by custom, cold.
> By custom, color.
> In truth – atoms and void. (fr. 9, fr. 125)

> But we, in reality, know nothing that is exact, but only what changes about according to the disposition of the body and of the entering and colliding ⟨*eidola*⟩. (fr. 9, from Sextus)

> Of knowledge there are two kinds, legitimate and bastard: of the bastard kind are all these – sight, hearing, smell, taste, and touch; the legitimate is different from these. (fr. 11, from Sextus)

> [The senses say] 'Miserable Mind, do you take from us your proofs and then cast us down? This casting down is *your* downfall.' (fr. 125, from Galen)

What is most obscure here is the manner in which the mind is supposed to operate according to Democritus' theory (it is a major problem in later Greek atomic theory too, as we shall see when we discuss the Epicureans). Perhaps the idea is that certain notions about the external world are suggested to the mind by the senses, and some of them (as opposed to others) cannot reasonably be doubted, or present themselves as somehow necessarily true. In the absence of further testimony about Democritus, it is illuminating to compare a passage from Galileo, aptly quoted in this same

[25] I take it that this is the interpretation suggested by Burkert, *ibid*. It must be emphasized again that it is an educated guess at the meaning of some very scrappy texts.

context by John M. Robinson (*Introduction to Early Greek Philosophy*, p. 202):

I feel myself impelled by the necessity, as soon as I conceive a piece of matter or corporeal substance, of conceiving that in its own nature it is bounded or figured in such and such a figure, that in relation to others it is large or small, that it is in this place or that place, in this or that time, that it is in motion or remains at rest, that it touches or does not touch another body, that it is single, few, or many; in short by no imagination can a body be separated from such conditions: but that it must be white or red, bitter or sweet, sounding or mute, or a pleasant or unpleasant odor, I do not perceive my mind forced to acknowledge it necessarily accompanied by such conditions; so if the senses were not the escorts, perhaps the reason or imagination by itself would never have arrived at them. Hence I think that these tastes, odors, colors, etc., on the side of the object in which they seem to exist are nothing else than mere names, but hold their residence solely in the sensitive body; so that if the animal were removed, every such quality would be abolished and annihilated. (*Il Saggiatore: Opere*, 1842 ed., vol. IV, p. 333)

Shape, size, relative position, and motion are the properties ascribed to the atoms themselves in Democritean theory. 'Tastes, odors, colors, etc.' are relegated to a lower order of existence. Whereas Galileo says in this passage that they are 'mere names,' Democritus says they are 'by custom.' Is it possible to get clearer about the meaning of this?

Democritus' position is to be distinguished from skepticism, in spite of the attempt by the skeptical Sextus – our only source for several fragments – to make a skeptic of him.[26] He did not claim that it is impossible to know anything, nor even that truth is not to be found in the reports of the senses. On the contrary, as Aristotle informs us more than once, it was his view that 'what is true is what appears'[27] – although Aristotle could also write:

[Some thinkers say,] furthermore, that we human beings, as well as many of the other animals, have opposite sense-impressions, and even one and the same individual does not always have the same impression from sense-perception. So it is unclear which set of these is true and which is false, since these are no more true than those, but equally so. So Democritus says either nothing is true, or to us at least it is unclear (*adelon*). (*Metaph.* IV.5, 1009b9–12)

If there is any consistency in these reports of Aristotle, he must mean that for Democritus the truth conveyed by sense-perception – although it *is* truth – is unclear to us. Perhaps 'unclear' is used in an almost technical sense: it is not immediately clear, but needs interpretation. To use Demo-

[26] Fr. 11, quoted above, distinguishes bastard from legitimate knowledge, but there is no reason to think he believed we are all condemned to be parents of nothing but bastards.

[27] τὸ φαινόμενον, *De anima* I.2, 404a27–28; cf. *De generatione et corruptione* I.2, 315b10.

critus' other metaphors, it is 'dark,' and needs the illumination of atomic theory before we can understand it, or alternatively it is 'bastard,' and needs to be legitimated by recognition of its proper parentage in atoms and void.

For all the confusion in the ontological statements reported from Democritus by our very incomplete sources, it is clear from Theophrastus' report in *De sensibus* 65 that he held there is a difference between what is sharp and what is sweet to the taste, and that this difference is grounded in the shape and arrangement of the component atoms. When he wrote 'by custom sweet, by custom bitter' and the rest, he must have meant by 'custom' something similar to Empedocles, who said (fr. 9) that he was following the 'custom' in speaking of 'coming-to-be' and 'ill-fated death.' There is, in Empedocles' theory, no real birth or death of mortal things, only mingling and separation of unchanging elements (fr. 8); but in expounding his own views he was willing enough to use these terms (frs. 17.3–4, 11, 35; 21.14; 26.4 and 10). The simple, customary term 'coming-to-be' is misleading, but can be used as a short way of referring to a complex feature of the real world. Similarly Democritus could say that although 'sweet' and 'bitter' do not label anything that is elementary and irreducible, nevertheless they make a distinction that is not arbitrary and unreal but based on a real distinction in the nature and arrangement of the component atoms.[28]

[28] This account of Democritus' position follows McKim, 'Democritus against Scepticism,' in *Proceedings of the 1st International Congress on Democritus*. See also Guthrie, *HGP*, vol. II, pp. 454–65, for a handy review of the evidence and a similar conclusion. Barnes, *The Presocratic Philosophers*, vol. II, pp. 257–62, makes Democritus a reluctant skeptic.

10 The cosmos of the Atomists

10.1 *The infinite universe*

It is particularly important, when writing about the Atomists, to be consistent in making the distinction between the universe or 'the all' (*to pan*, in Greek) and the world (*kosmos*) – that is, a particular part of the universe consisting of our earth, sea, air, and visible stars, planets, moon, and sun. Leucippus and Democritus held that 'the all' consists of an unlimited expanse of void space and an unlimited number of atoms. Of these, some are – temporarily – arranged so as to form worlds, including the world in which we live. We must study, first, the arguments used to defend the notion of the unlimitedness of the universe, and then the mechanisms that were suggested to account for the formation of worlds.

If we are to understand the controversy of classical antiquity, we must keep clearly in mind one important difference between ancient and modern cosmology. In the twentieth century we have grown accustomed to the notion that what we see in the sky on a starry night is the beginning of a universe that continues far beyond the limits of our vision. We can see some of the stars, because they are relatively large, bright, and near; with telescopes we can see further into the distance and see more stars, and with better telescopes we could see further still. In the cosmology of classical antiquity, on the other hand, it was a matter of common agreement that the stars we see are part of our world: they are the boundary *beyond which* the infinite universe (if it is infinite) begins. Both the Atomists, who believed in the infinite universe, and the Aristotelians, who did not, agreed that our world is itself a finite system, bounded by the sphere of the stars. The controversy was about what, if anything, lies beyond the starry sphere.[1]

As so often, for information about the Atomists' theory we must go to a hostile witness: Aristotle. In his *Physics* he sets out five reasons that have led people to believe in 'the infinite.'[2] He mentions no names, and the at-

[1] For more on this subject, see my article, *Journal of the History of Ideas* 42 (1981), to be reprinted in my *Cosmic Problems*.

[2] III.4, 203b15–30. The following section draws on my article, 'Aristotle and the Atomists on Infinity,' in Düring, *Naturphilosophie bei Aristoteles und Theophrast*, to be reprinted in my *Cosmic Problems*.

tribution to Democritus must be judged in each case on its merits. The conviction that the infinite exists rests on (A) the infinity of time and (B) the infinity of magnitudes; (C) the belief that genesis can be kept up only if there is an infinite store to draw on; (D) the conception that a thing can be limited only if there is something beyond it; and (E) the fact that one can always imagine something further in the number series, in geometrical magnitudes, or even outside the cosmos.

(D) and (E) are the important ones in the present context. The notion of (D) seems to be as old as the fifth-century Eleatic, Melissus. From the premiss that what *is* is infinite, he drew the conclusion that it must be one, since if there were two beings they would limit each other.[3] It may be that he also used the converse entailment, that if a being is limited, it must have something other than itself to limit it.[4] Aristotle criticizes the argument on the ground that there is a distinction to be drawn between *limit* and *contact*: it is true that contact must be between two things, but a thing is said to be limited by reference to itself alone.[5] Thus he defends his own view that the cosmos is finite and has nothing – literally *nothing*, not even empty space – outside it.

A form of argument (E) is attributed to the Pythagorean Archytas, a contemporary of Plato.[6] If one stood on the outermost edge of the universe (supposing that it had one), could one stretch out one's hand beyond the edge, or not? It is absurd to imagine that one could not, and so it appears that there must be at least *place* beyond the edge. Centuries later, Lucretius uses basically the same argument, substituting a javelin for the outstretched hand.[7] The case for attributing it to Democritus rests on its use by his fellow-Atomist Lucretius and by his near-contemporary Archytas, and on Aristotle's mention of the void in his development of the argument. Aristotle himself objects to it, in effect, that such a thought-experiment cannot prove anything about the facts: although one can always *imagine* something beyond the cosmos, that does not prove that there *is* something there. This objection hardly satisfies one's intuitive sympathy with Archytas, however, and even so good an Aristotelian as Simplicius was worried by his argument.[8] It is a more powerful argument than (D), since it rests on a common intuition about the nature of space, whereas Aristotle appears to be right, in his criticism of (D), that it is poss-

[3] Melissus fr. 5. See also Ps. -Aristotle, *De Melisso* 974a11.

[4] This seems a reasonable inference from Aristotle, *De generatione et corruptione* I.8, 325a15.

[5] *Physics*, III.8, 208a10–14. Epicurus reaffirms the argument in the face of Aristotle's criticism in *Letter to Herodotus* 41.

[6] Eudemus, quoted by Simplicius, *Ph.* 467.26ff.

[7] Lucretius, *De rerum natura* I.968–79.

[8] Simplicius, *Physics* 467.35.

ible to give an account of *limit* without considering anything beyond the limit.

Having shown that the void is infinite, the Atomists argued that the number of atoms must be infinite, too. Aristotle quotes an argument (C) that at first sight appears to be anachronistic for the Atomists. As we have seen, they were persuaded by Parmenides that nothing comes into being or perishes, in the last analysis: hence the eternity of genesis can be maintained in cycles of rearrangement of eternal material constituents; and one would suppose that a finite stock of atoms could, in principle, account for this. Whereas the argument might appeal to someone before Parmenides – Anaximander, perhaps – it could hardly be thought cogent by any post-Parmenidean. But there may be more to it than that. If a finite stock of atoms were dispersed in an infinite void space, they might never meet at all, and so never combine into a compound. This is the form the argument is given by the Epicureans,[9] and it may well be the original Democritean form, too.

However, there is a more interesting development of the Atomists' case. After mentioning argument (E), Aristotle continues:

> [E1] The exterior region [*sc.* the region outside the boundaries of the cosmos] being infinite, it seems that body, also, is infinite, and worlds, too; for why *here* in the void, rather than *there*? So if in one place, there should be a mass in every place. [E2] Moreover, if the void and place are infinite, body must be too; for in eternal things there is no difference between *can be* and *is*. (*Physics* III.4, 203b25–30)

If we leave on one side for the moment the brief mention of the notion of infinite worlds, the most interesting point here concerns the style of argument. E1 depends on some notion of a *principle of sufficient reason*. The argument is that since we see a certain concentration of matter in the portion of the universe that houses our world, and since the universe as a whole – that is to say, the void – is everywhere invariant in quality, so that it is impossible to imagine anything that singles out our region as *special*, it follows that we must imagine every region of the universe to be similar to ours in housing about the same amount of matter. There is no sufficient reason for ours to be different.

An argument of the same kind had been used by Parmenides, as part of his polemic against genesis.

> What necessity could have forced it to grow, starting from nothing, *later or sooner*? (fr. 8.9–10)

[9] Epicurus, *Letter to Herodotus* 41–2; Lucretius, *De rerum natura* I.1014–20; Diogenes of Oenoanda, fr. 6.II.

This is an argument that starts from the invariancy of *time*. Since all times are similar, it is impossible to think of a sufficient reason for the growth of a world starting at one time rather than any other. Hence we are not to think of such a growth at all. Democritus argues from the invariancy of *the void*. He cannot conclude, as Parmenides did, that no world ever grew, since he is committed to the proposition that our world, at least, did grow. So his conclusion is that other regions of the void must also contain worlds.

The principle 'not rather this than that' is comparatively well attested for Democritus. We have already seen the expression used in connection with the existence of void: 'not rather the thing than the not-hing.' Simplicius reports that he attributed an unlimited variety of shapes to the atoms on the ground that 'nothing is rather of this kind than of that' (*Physics* 28.25–6). Theophrastus mentions his use of the principle to support the idea that no sense-perception has a better claim to reach the truth than any other (*De sensibus* 69).

E2 appears to depend on the premiss that a place is to be defined as something that can receive a body; so at least Philoponus understood it (*Physics* 408.10–16). If there is an infinite extent of place, there must be an infinite extent of body, because a finite quantity of it would soon be exhausted, leaving parts of space unfillable. This seems a very odd argument, since it could easily be circumvented by supposing that all of space is fillable by a finite quantity of body in motion, given that time is also infinite. The 'principle of plenitude' invoked by Aristotle – that all possibilities are actualized in infinite time – does not belong to Democritus, as far as I can see. So it may be that Democritus added the infinity of body to the infinity of space on the ground of the intuition that his universe must be invariant everywhere. If space were infinite but body finite, then there would be some regions of the universe containing body and some not: 'but why here rather than there?' Alternatively, we might guess that he used the same argument that the Epicureans were to use, that a finite number of atoms scattered in an infinite void would never meet, and so never form a cosmos at all (Lucretius 1.1017–20).

The existence of infinite worlds is certainly a Democritean thesis. If the view we have taken of Anaximander and Anaxagoras is correct, Leucippus and Democritus were the first to advance this idea. It is an idea that seems to present an invitation to let the imagination loose in a speculative canter – especially in view of such contemporary fantasies as Aristophanes' Cloudcuckooland and the somewhat later Atlantis myth. It is disappointing that the surviving evidence tells us so little about this feature of the Atomists' theory. All we have is a tantalizing summary:

Worlds are infinite [in number] and different in size. In some there is no sun or moon, in some they are larger than ours, and in some more numerous. The intervals of the worlds [i.e. the distances between them?] are unequal; in one region there are more, in another fewer. Some are growing, some in their prime, some waning; here they come to be, there they fail. They perish by colliding with each other. Some worlds are barren of animals and plants and all moisture. (Hippolytus, *Refutations* 1.13.2=DK 68A40)

The feature of the worlds that is stressed by most of the ancient authors who mention them is their transience. After Aristotle had argued in his famous dialogue *On Philosophy*, now lost, and in his treatise *De caelo*, that our world is eternal, without beginning or end, this came to be one of the chief debating-points between the schools. As we have explained in chapter 1, the fifth-century Atomists and the Epicureans represent the opposite extreme to Aristotle, with various factions – for example, the Stoics – occupying the intermediate ground.

10.2 *The origin of our world*

Our world, like all the other worlds postulated by the Atomists, had an origin at some particular time and will perish in due course. Its origin is explained by natural causes, without the aid of any supernatural planning agency: everything comes about 'by necessity,' as a result of matter in motion in the infinite void. This is the doctrine that we must now analyze as closely as the evidence will permit.

The best narrative of the Atomists' cosmogony is given by Diogenes Laertius, who attributes the following to Leucippus:

The coming to be of the *kosmoi* is thus. (1) In severance from the infinite, many bodies, of all varieties of shape, move into a great void. (2) These, being assembled, create a single vortex [δίνη], in which they collide, gyrate in every way, and are sorted like to like. (3) When because of the number they are no longer able to move round in equilibrium, then the fine ones move into the void outside, as if sifted, while the remainder stay together, become intertwined, join courses with each other, and bring about a first system, in the shape of a sphere. (4) A sort of membrane comes apart from this,[10] containing in itself bodies of all kinds. (5) As these bodies whirl around in proportion to the resistance of the center, the surrounding membrane becomes thin (fine), as the bodies that are contiguous by their contact in the whirl continually flow together, and thus the earth comes about, as the bodies that are carried to the center remain together. (6) And again the containing membrane grows bigger by influx of external bodies: being moved in a whirl, it seizes upon whatever it touches. (7) Some of these bodies, being tangled together, form a system, at first moist and muddy, but later they are dried

[10] Adopting the reading τούτου, for MS τοῦτο, with Kerschensteiner, *Hermes* 87 (1959).

out and carried around with the whirl of the whole, and then they are ignited and form the nature of the stars.

(8) The circle of the sun is outermost, that of the moon nearest to the earth; the others intermediate. (9) And all of the stars are ignited by the speed of their motion but the sun is also ignited by the stars (?); the moon, however, has little of fire. (10) The sun and moon are eclipsed ... (11) ... by the earth's being tilted towards the south. The regions toward the north are always snowy and cold and frozen. (12) The sun is seldom eclipsed, the moon frequently, because of the inequality of their circles.

(13) As is the genesis of a *kosmos*, so is its growth, waning and destruction, by some necessity, the nature of which he [Leucippus] does not elucidate.[11] (Diogenes Laertius IX.31 = DK 67A1)

The origin of the vortex is described in sentences (1) and (2), but not explained. We shall return in the next section to consider the Atomists' theory of motion; for the present, we shall merely note that sentence (13) gives us some authority for saying that the origin was ascribed to 'necessity.' The first move in cosmogony involves a 'severance from the infinite' and a move 'into the great void.' The former phrase is not clear, but I am inclined to think that it means no more than that a certain quantity of atoms from the infinite stock,[12] not being implicated in any complex, moves into a region of the void formerly empty of all matter. This is a very puzzling clause. The atoms are moving in the infinite void already: why therefore do they need to move 'into a great void' before a cosmos is formed? I think the answer may be simply that the observed form of the world suggests that it contains a very large space, between the earth and the stars. Twentieth-century cosmology has accustomed us to the idea that the distance between the earth and the moon, sun, and planets is very small compared with the distance between the sun and other stars – in other words, that the solar system is a small unit in huge volumes of empty space. But this is a recent development: even Kepler thought of the world rather differently:

Be it admitted as a principle that the fixed stars extend themselves *in infinitum*. Nevertheless it is a fact that in their innermost bosom there will be *an immense cavity*, distinct and different in its proportions from the spaces that are between the fixed stars. So that if it occurred to somebody to examine only this cavity ... from the sole comparison of this void with the surrounding spherical region, filled

[11] I have inserted the reference numbers. There is an excellent analysis of this text by Kerschensteiner, *ibid*.

[12] The article with ἀπείρου is feminine. Diels and others take this to refer to infinite space (χώρα), but it seems to make better sense as 'the infinite nature' (φύσις). ἀποτομὴ ἀπὸ τοῦ ἀπείρου occurs in Epicurus, *Letter to Pythocles* 88, so there may be some contamination here.

with stars, he would certainly be obliged to conclude that this is a certain particular place and the main cavity of the world.[13]

It may be this same intuition, that the hollow between the stars and the earth is a *special* condition in the universe, that led the Greek Atomists to postulate a 'great void' for the formation of each one of the worlds.[14]

The vortex is already familiar from the work of Anaxagoras and Empedocles (see above, sections 6.2 and 7.3). Anaxagoras' vortex and its sorting power were ascribed to the agency of Mind, and Love and Strife were involved in the process in Empedocles' theory. The atomistic version, on the other hand, mentions no motive agencies at all: the vortex simply occurs, and as a result, the bodies involved in it 'are sorted like to like.' The vortex does not take place in a medium that is distinct from the bodies affected by it, as in later theories; the atoms themselves constitute the vortex by forming among themselves a circular pattern of motion. We shall return in the next section to consider the theory of motion that lies behind this, but we should observe in passing that the sorting of like to like is conceived as a purely mechanical process. There is no place in the theory for any kind of attractive force, acting between bodies at a distance, or even between bodies in contact. Atoms of similar size and shape tend to assemble in the same region of the vortex just because they must behave similarly in similar conditions: once assembled together, they stay together by becoming interlocked with each other. The kind of mechanical sorting that is postulated in this process is illustrated in a fragment of Democritus' own words preserved by Sextus Empiricus:

> Living creatures flock together with creatures of the same kind, as doves with doves and cranes with cranes and so on – thus too in the case of inanimate things, as you may see with seed that is sieved and pebbles on the seashore: in the one case, as the sieve is moved around [the verb, δῖνον, is cognate with the word for 'vortex,' δίνη] lentils are sorted and ranged with lentils, barley with barley, and wheat with wheat; in the other case, as the waves move, long pebbles are pushed into the same place with long pebbles, and round with round, as if the likeness in these things had some [force] that tended to collect them together. (Democritus, fr. 164)

The qualification 'as if' in the last clause is important.

Sentence (3) in the account of the Atomists' cosmogony presents some

[13] *De stella nova*, p. 689, quoted by Koyré, *From the Closed World to the Infinite Universe*, p. 62.

[14] It should be noted, however, that the Atomists' reasons for this view would not be the same as Kepler's. His reasoning, based on the observed size of the stars, was that from any given star the stars nearest to it would appear much larger than any star appears from earth. Hence the earth is in a much less densely populated region than the rest of the universe. When the Greek Atomists thought of the universe outside the world, they thought of totally unseen regions *beyond* the stars.

problems. What is the equilibrium that is mentioned? The word itself suggests a state in which there is no tendency to move in one direction in preference to another, and this is sometimes related to the image of the random motions of dust particles, illuminated by a ray of sunlight – an image attributed to Democritus by Aristotle.[15] However, the atoms are said to be 'no longer able to *move around* in equilibrium,' and that seems fatal to the idea that the phrase refers to the period before the vortex was formed. Probably the idea is that large and small atoms are whirled around *together* until crowding causes the smaller ones to be pushed toward the outside. The atoms that remain in the vortex are said to form 'a first system, in the shape of a sphere.' The spherical shape is, to my mind, the greatest puzzle in this narrative. Something was said in earlier chapters about the difficulties of fitting the vortex model to a spherical cosmology: a vortex is a system of motion around a *linear* axis, and in principle there seems to be no reason why it should assume the shape of a sphere. The collection of bulkier matter in the center is naturally thought of as being at the *bottom* of the linear axis, and we have seen that Empedocles put forward a special explanation of how the earth is supported some way above the bottom. The Atomists believed the earth to be disk-shaped, and probably thought of it as supported on air.[16] These two theses show that the Atomists subscribed to the linear theory of motion, not to the centrifocal theory of Aristotelian cosmology. We must return to this most important topic in the next section.

Sentences 4 and 5 of our cosmogonical text describe first the formation of a membrane or caul around the outside of the embryo cosmos, and then the sorting of atoms within the enclosed collection thus formed. It is illuminating to compare this description, sketchy as it is, with some passages from the little treatise on embryology that was attributed in antiquity to Hippocrates, although it probably dates from the fourth century B.C. It is worth quoting for many reasons, but perhaps especially because it is useful to be reminded that this is too early in the history of cosmology to make a sharp distinction between a mechanical and a biological model.

'Hippocrates' describes the first stages as the mixing of the seed of both parents in the womb, the condensation of this mass through heat, and its acquisition of a kind of breathing process. He then continues:

[15] Aristotle, *De anima* I.2, 403a28, connects the image with movements of soul atoms. Burnet, *Early Greek Philosophy*, p. 365, suggests that it is to be applied to pre-cosmic motions.

[16] Leucippus is said to have made it 'drum-shaped' (DK 67A26). Democritus, according to one report, made it not circular but oval, its length being 3/2 times its breadth (DK 68B15); 'disk-shaped,' on the other hand, according to Aetius (DK 68A94).

As it inflates, the seed forms a membrane around itself; for its surface, because of its viscosity, stretches around it without a break, in just the same way as a thin membrane is formed on the surface of bread when it is being baked; the bread rises as it grows warm and inflates, and as it is inflated, so the membraneous surface forms. In the case of the seed, as it becomes heated and inflated the membrane forms over the whole of its surface, but the surface is perforated in the middle to allow the entrance and exit of air. In this part of the membrane there is a small projection, where the amount of seed inside is very small; apart from this projection the seed in its membrane is spherical.[17]

The author claims to have seen a six-day embryo himself when he caused a dancing-girl to abort by getting her to jump vigorously up and down, touching her buttocks with her heels.

He continues with his theory. The embryo draws in blood, along with breath, and this blood coagulates and begins to form flesh.

As the flesh grows it is formed into distinct members by breath. Each thing in it goes to its similar – the dense to the dense, the rare to the rare, and the fluid to the fluid. Each settles in its appropriate place, corresponding to the part from which it came and to which it is akin. I mean that those parts which came from a dense part in the parent body are themselves dense, while those from a fluid part are fluid, and so with all the other parts: they all obey the same formula in the process of growth. The bones grow hard as a result of the coagulating action of heat; moreover they send out branches like a tree. Both the internal and external parts of the body now become more distinctly articulated... Now the formation of each of these parts occurs through respiration – that is to say, they become filled with air and separate, according to their various affinities. Suppose you were to tie a bladder onto the end of a pipe, and insert through the pipe earth, sand and fine filings of lead. Now pour in water, and blow through the pipe. First of all the ingredients will be thoroughly mixed up with the water, but after you have blown for a time, the lead will move toward the lead, the sand toward the sand, and the earth toward the earth. Now allow the ingredients to dry out and examine them by cutting around the bladder: you will find that like ingredients have gone to join like. Now the seed, or rather the flesh, is separated into members by precisely the same process, with like going to join like. So much, then, on that subject.[18]

It seems likely to me that sentence 6 in our text about the birth of the cosmos, which is otherwise rather mysterious, can be explained by analogy with the embryological text. The membrane that surrounds the cosmos grows bigger by influx of external bodies. This is exactly what happens to the human embryo, according to 'Hippocrates': the seed, contained in a membrane, grows because of its mother's blood, which descends to the womb (*The Nature of the Child* 14). Leucippus may well have been motivated to include this stage in his cosmogony for no other reason than to make it conform to current embryological theory.

[17] Hippocrates, *The Nature of the Child* 12, trans. Lonie.
[18] *Ibid.* 17.

On no account must this suggestion of extra-cosmic material be allowed to confuse our picture of the essentially monistic theory of matter held by the Atomists. The next sentence, no. 7, tells us that some of this material becomes the heavenly bodies; but fortunately it makes clear beyond doubt that this is no special kind of material. For one thing, only part of it ('some of these bodies') is taken up into the system of the heavens; the rest, we may presume, forms the body of the lower cosmos. If my suggestion in the last paragraph is correct, the material drawn in from outside must feed the whole body of the cosmos, not just the stars. In any case, the matter of the heavens is plainly described as having been 'moist and muddy,' and later dried out and ignited. This is nothing like the Aristotelian aether or 'first body' with its unique and special properties, and it is very important to keep this clear. That radical dualism between heaven and earth that was still bedeviling science when Galileo wrote his *Dialogue concerning the Two Chief World Systems* was no part of the Atomists' theory.

There is confusion in the reports of the details of the Atomists' astronomy. If Leucippus held that the sun was the outermost of the heavenly bodies, as one text says, he was apparently corrected by Democritus, who put the fixed stars on the outside, then the planets or some of them, then the sun; he agreed with Leucippus (and with virtually all Greek astronomers) that the moon is closest to the earth.[19]

The true explanation of eclipses had been worked out some decades before Leucippus and Democritus, as we have seen, and there is evidence that Democritus, at least, believed that the moon shines with light reflected from the sun.[20] For this reason, editors have emended the text of Diogenes, which says, according to the manuscript, that the tilt of the earth toward the south is the cause of eclipses. The tilt of the earth relative to the north pole was introduced, as we have seen, to account for the fact that the star-circles are not parallel to the surface of the earth, as they should have been on the vortex-theory of the origin of the cosmos.[21] It is difficult to see how it could have anything to do with eclipses, at least of the sun, and it is therefore fairly safe to suppose that the text is somehow wrong. Another source offers a physical reason for the tilting of the earth: the 'surrounding' (i.e. the supporting air) in the south is mixed (with what?) whereas in the north it is unmixed and therefore stronger.[22] So it is 'weighed down' in the south, where there is a greater quantity of growth

[19] Diogenes Laertius 9.33 = DK 67A1. Aetius II.15.3 = DK 68A86.
[20] Plutarch, *On the Face in the Moon* 16, 929C.
[21] See above, p. 93 note 13 and accompanying text.
[22] Aetius III.12.2 = DK 68A96.

and vegetation. This information could perhaps be put together with sentence 11 of the Diogenes text without inconsistency, if we suppose that the frozen north in that text and the lush vegetation of the south are consequences of the tilting, not causes of it. The earth is held to be somewhat dished in the middle of its disk: so it is plausible to hold that the north, being tilted up, is more shadowed than the south, and therefore more chilly and less productive. But this is speculation, on shaky evidence.

10.3 *The Atomists' theory of motion*

The confusion and incompleteness of the evidence on this subject is lamentable. There is still no agreement among scholars about many dubious points in the Atomists' theory. It is therefore all the more important to be clear about the undisputed parts. These are sufficient, I believe, to show us that Leucippus and Democritus launched an enterprise of the utmost importance in the history of science.

What is crucial is that they produced a *unified* theory. We have already seen that Anaxagoras and Empedocles, spurred by the challenge of Parmenides, were the first for whom we have evidence of an explicit theory of motion. Both of them were dualists. Anaxagoras made a distinction between the substances of the physical world on the one hand, and Mind on the other: the distinction was not between material and immaterial beings, but between passive things and the actively organizing Mind. Empedocles similarly made a division between the four material elements and the two moving forces, Love and Strife. The Atomists dispensed with all such animistic agencies, and claimed that the world and its workings could be explained by nothing but the collisions of atoms, which move until they collide, since there is nothing to stop them, in the void. Zeus the Cloud-gatherer, Poseidon the Earth-shaker, all the cosmic gods of mythology, and their counterparts in early cosmological theories, were thus released from their jobs.

Leucippus and Democritus avoided another kind of dualism, too. It was pointed out in the last section that in their theory there is no basic difference between the matter of the heavens and the earthly elements. The same is true of their motions. Aristotle, faced with the fact that the heavenly bodies move in circular orbits whereas the material substances on and around the earth's surface fall or rise in straight lines, posited two different kinds of matter to correspond with the two different kinds of motion. And this dualism was to prove singularly long-lasting, as we have mentioned already. The Atomists avoided it. The model of the vortex satisfied them: it showed how the formation of a cosmos might get started without a Mind to initiate it, and how the different elements of the cosmos might get sorted out without any marriages arranged by Love or divorces brought about by Strife.

We must look at some of the evidence, difficult as it is. Much of the difficulty arises, as we have mentioned before, from the fact that our best sources are critics and opponents of Atomism. They are not bent on giving an unbiased or systematic account of the theory they reject.

Aristotle says:

> There are some who suppose that spontaneity (τὸ αὐτόματον) is the cause of this heaven and of all the *kosmoi*. From spontaneity, they say, comes the vortex, that is, the motion that sorted out and established the universe into its present order. (*Physics* II.4, 196a24–8)

Although Aristotle gives no names here, Simplicius mentions Democritus among others as fitting the description in that he says that a vortex, of all kinds of shapes (of atoms), was separated off from the universe but does not specify how or by what cause (*Physics* 327.23–5). The mention of plural worlds by Aristotle confirms that he had the Atomists in mind.

What does Aristotle mean by τὸ αὐτόματον? It is a technical term in his philosophy: 'among things that generally speaking come about for the sake of something, whenever things whose cause is external come about *without* doing so for the sake of the outcome, we say they come about from τὸ αὐτόματον' (*Physics* II.6, 197b18–20). We must look at this in context in volume 2; for the moment a rough idea will suffice. Aristotle first picks out a type of sequence of events, which he calls 'things that generally speaking come about for the sake of something,' such that we recognize the outcome as being the end or goal of the process that led up to it. Examples could be found abundantly in plant life: we would recognize the process of budding and fertilization of fruit trees as being *for the sake of* the fruit, or more generally for the sake of the continuity of the species by means of the seed. Sometimes, however, such a sequence can be identified when we are not prepared to say that the process was for the sake of the outcome. A storm, for example, might so reshape a stretch of the shore that what had previously been a sandy shallow becomes a haven for breeding oysters. The key points in this example are that it is a natural, not an artificial, process; that its outcome is somehow good – for oysters, if not for man; and that it is not a frequent or normal outcome but recognizable as a kind of freak.

According to Aristotle, the Atomists believed the world to have originated as a freak of this kind. There is, as it were, a storm among the atoms, and some happen eventually to be flung into the pattern that constitutes a cosmos. It is *as if* planned, but it was not planned.

We shall see later that Aristotle's description is a loaded one: he means it to be the beginning of a criticism. But we have to use his evidence for the Atomists' theory, because there is so little else. What we can infer from it with some certainty is the negative point: the Atomists gave no explanation for the formation of a cosmos beyond a description of the atoms and a statement that a vortex arose among them, so that they were sorted out by shapes and sizes.

The assertion that worlds originate by spontaneity (τὸ αὐτόματον) in the Atomists' theory is sometimes coupled with the statement that they arise from *necessity*. A fragment of Epicurus apparently refers to Democritus as attributing the cause of everything to 'necessity and spontaneity.'[23] Another text reports that according to Democritus, 'all things happen by necessity, the cause of coming-to-be being the vortex, which he calls necessity.'[24] Aristotle himself could substitute necessity for *to automaton* in a statement about Democritus' theory of causation very similar to that just quoted from the *Physics*.[25]

How does it come about that necessity can be coupled in this way with τὸ αὐτόματον, or spontaneity?[26] That is a question that cannot be discussed adequately until we have studied the contrast drawn between necessity and mind in Plato's *Timaeus*, and Aristotle's contrast between necessity and purposiveness in his physical and biological works. We shall see then how 'necessity' came to be associated especially with processes brought about by matter in motion in abstraction from any directiveness imparted by complex forms or by intelligence. Once these distinctions had been made, it became a matter of fierce controversy whether 'necessity' by itself would ever account for the evident regularity and directiveness of nature. But Leucippus and Democritus, we assume, walked in all innocence into this future battlefield. Leucippus could write (it is the only direct quotation of his that survives): 'Not one thing comes to be randomly, but all things from reason and by necessity.'[27] A Platonist, if he said something like that, would mean that these two kinds of cause *taken together* can account for everything: everything is *either* from reason *or* by necessity. Leucippus apparently drew no contrast between them, but meant simply that for everything there is a reason, namely that it had to be

[23] τὸ αὐτόματον: Epicurus, fr. 34.30, Arrighetti, *Epicuro*, 2nd ed.
[24] Diogenes Laertius IX.45 = DK 68A1. Cf. also Sextus IX.113 = DK 68A83.
[25] *De generatione animalium* V.8, 789b2.
[26] See Edmunds, *Phoenix* 26 (1972), and Sorabji, *Necessity, Cause and Blame*, pp. 17–18. Balme's articles, *CQ* 33 (1939) and *CQ* 35 (1941), were pioneers in this area.
[27] DK 67B2.

so, given the state of the atoms and their motions. Democritus was said by Eudemus to have done away with chance (τύχη) by referring every event that might be attributed to chance to other causes. Thus the cause of a 'lucky' find of buried treasure was digging or planting an olive tree; the cause of the 'unlucky' death of the bald man in the story was that the eagle dropped the tortoise, hoping to break its shell.[28]

The major theme of Aristotle's complaint is that the Atomists gave no explanation of motion. 'Concerning motion,' he says, 'whence and how it comes to be a property of things, these people ... lazily left the question on one side' (*Metaphysics* I.4, 985b19–20). The commentator Alexander remarks about this: 'Leucippus and Democritus say that the atoms move by colliding with each other and crashing against each other, but they do not say whence comes the principle of their natural motion' (36.21). It is a charge Aristotle returns to again and again: they say the atoms are always moving, but they do not say why, or what their natural motion is (*De caelo* III.2, 300b8; *Metaphysics* XII.6, 1071b31).

Clearly the Atomists claimed that atoms were always colliding with each other, and that this had always been so and always will be so. Aristotle was wedded to his own theory, which made a distinction between forced motion, like that of a stone thrown upwards, and natural motion, like the fall of a stone when it is dropped. His objection is that one has no right to posit forced motions without any natural motion to contrast them with. The Atomists would have replied in this way. Two atoms collide because they are on a collision course through the void. The effect of the collision on their subsequent motion will depend on their shape, size, and weight: perhaps also on their speed and their angle of impact. These factors are a sufficient explanation of the atoms' movement after the collision. If you ask how they came to be on a collision course in the first place, the answer is that their motions were due to previous collisions with other atoms. And this answer can be repeated *ad infinitum*, because there are no spatial or temporal boundaries to the universe. Atoms have been colliding for ever.

We should be careful not to ascribe too much to Leucippus and Democritus. It might sound as if they anticipated Newtonian laws of motion, and claimed that an atom will continue to move through the void with no change of direction or velocity until impeded by collision with another atom. I doubt if they had any such precise idea of the motion of atoms. Some of the few scraps of evidence we have use the verb

[28] Simplicius, *Physics* 330.14–20, commenting on Aristotle, *Physics* II.4, 196a14. The unlucky bald man is sometimes said to have been Aeschylus.

περιπαλάσσεσθαι, meaning 'to be sprinkled about.'[29] Aristotle says Democritus compared the atoms composing fire and psyche to the motes of dust we can see dancing in the rays of sunlight shining through a window.[30] Skeptical scholarly eyes have found reasons to be suspicious of this evidence, but I think it may well be genuine. At least it offers a plausible picture of the Atomists' ideas. They were not interested enough to speculate about the 'free' movement of an atom in the void. Perhaps no atoms were 'free' in this sense: they were crowded, so that they collided with each other, and the only motions worth speculating about were those of atoms in more or less dense crowds. In such conditions they could provide the basis of an explanation of a vortex with its capacity for sorting by size, shape, and weight, and of the upward movement and suspension of basically heavy particles. A fundamental thesis of the Atomists involves the notion of something being squeezed out by pressure (ἐκθλίβεσθαι): the dense crowding of atoms thus explains how some are forced upwards. Aristotle complains that this, along with being left behind by atoms falling more rapidly, is the only explanation offered by the Atomists for upward movement: they had no idea of elements with a natural upward movement.[31]

What is the role of weight in Democritus' atomic theory? The evidence is contradictory, and the answer is extremely controversial. Some hold that weight was not a fundamental property of the Democritean atom at all, but only something derived from the forces of the vortex.[32] A recent interpretation states that weight was indeed a property of all atoms, but was not manifested in a tendency to downward motion except in the special conditions of a vortex.[33] My own view – a revival of an old one – is that Democritus always thought of atoms as being more or less heavy, and therefore having more or less of a tendency to move downwards.[34] Aristotle's frequent complaints that he had no theory of natural motion may be explained, I believe, on the grounds that Democritus never distinguished between weight and collision as being forced and natural respectively, and never thought of one as 'original' and the other as secondary: atoms had weight always, and were *always* colliding and changing

[29] Leucippus, DK A15 = *De caelo* III.4, 303a7; Democritus, DK A58 = Simplicius, *Physics* 1318.35 = DK B168; Democritus, DK A135 = Theophrastus, *De sensibus* 66. But McDiarmid, *Hermes* 86 (1958), casts all these in doubt.

[30] *De anima* I.2, 404a1–5.

[31] Aristotle, *De caelo* IV.2, 310a10.

[32] For example, Bailey, *The Greek Atomists*, pp. 129–32, 144–6. He is followed by most subsequent writers.

[33] O'Brien, *Theories of Weight*, vol. I: 'Democritus: Weight and Size.'

[34] For more, see my review article on Mr O'Brien's book mentioned in the last note: *Oxford Studies in Ancient Philosophy* I (1983), to be reprinted in my *Cosmic Problems*.

their paths. Like Epicurus later, Democritus was bound to deny that the infinite void has a top or a bottom or a center, or any co-ordinates by which we can establish direction: directionality is a property of matter itself, not of the void. Unlike Epicurus, Democritus had no idea of a theoretical downward 'rain' of atoms, from which they had to be rescued by a swerve: every collision can be explained by previous collisions, *ad infinitum*, and there is no need, he thought, for any explanation of the set of collisions as a whole.

But the matter of weight cannot be explored at more length here. It is regrettable that it is not possible to be sure about it, but that is the case, and we have to accept it. This doubt, although it is a major one, still leaves our overall picture of the Atomists' theory intact.

Leucippus and Democritus were no doubt more interested in the behavior of atoms in the formation and dissolution of compounds than they were in the hypothetical case of atoms moving freely in space. But even on this subject the evidence is meager. It is clear that they used the shape, size, and weight of atoms as explanatory factors.

> The enduring coherence of the ⟨elementary⟩ substances with each other Democritus explains by the dovetailing and jointing of the ⟨primary⟩ bodies. Some of them are crooked, some hook-shaped, some concave, some convex, and so on with innumerable different shapes. So he thinks that they cling to each other and cohere just so long as no stronger necessity, arising from the environment, seizes on to them and scatters them apart. (Aristotle, *On Democritus*, in Simplicius, *De caelo* 295.14–20)

Elsewhere Aristotle tells us that in Atomist theory atoms differ from each other in three respects, and he gives their own words, with an explanatory gloss: ῥυσμός or shape, διαθιγή or order, and τροπή or position.[35] He gives an illustration: A differs from N in shape, AN from NA in order, I from H in position. These may be the Atomists' original illustrations:[36] the parallel between the relation of letters to syllables and that of elements to compounds is used by Lucretius (1.823–6), and indeed it seems an obvious one for an Atomist to draw.

[35] *Metaphysics* 1.4, 985b15–22.

[36] So Ross, *Aristotle's Metaphysics, ad loc.* But if so, Eudemus must have been wrong in saying Plato first gave the name *stoicheia* (letters) to elements (fr. 31, Wehrli).

11 The anthropology of the Atomists

One of the finest products of the Greco-Roman imagination is the fifth book of Lucretius' poem *De rerum natura*. In magnificent language the Latin poet pictures the successive stages of the evolution of the cosmos from clouds of atomic dust, and with great intensity displays the explanatory advantages of his materialism over the cosmic gods of ancient mythology and rival philosophies. First comes the framework of the cosmos: earth and sun, the stars, planets, sun and moon, and the seasons of the earth. Then the poem goes on to tell of the first growth of vegetation, the emergence of animal life, the earliest human communities, the growth of civilization and technology, the development of political structures, and the genealogy of morals.

There can be little doubt that the Atomists of the fifth century B.C. wrote about these things too, and that many of their ideas were adopted by Epicurus and his Greek followers, who were themselves the sources of Lucretius' inspiration. It happens that Lucretius' poem has survived intact, while the work of Leucippus and Democritus has vanished. The fifth book of *De rerum natura*, then, is our best source for this aspect of ancient Atomism. Nevertheless, it seems appropriate to study Lucretius in his Epicurean context, rather than as an appendix to Democritus. Although there is in all probability much in Lucretius book V that comes from Democritus, there is certainly much that does not. So at the cost of leaving our history of Preplatonic Atomism embarrassingly threadbare in this area, we shall not draw on Lucretius' anthropology for evidence here, and postpone the attempt to give a more satisfactory account of the atomic theory to volume 2. We are left, as usual, with exasperating scraps of disconnected information.

11.1 *The materialist theory of the psyche*

The debate as to whether the soul is an immaterial thing or a material one first emerges into the open, so far as we are concerned, with Plato's *Phaedo*, and it may well be that Plato was the first to put the issue in these terms.[1] There are other issues in ancient theories about the psyche with

[1] Especially *Phaedo* 78b–81d.

which this should not be confused. There is, for example, the question of immortality. Among later philosophers, on the whole, if the psyche was held to be material, it was also thought to be mortal, but this is by no means the case in the earliest texts. Then there is a different distinction, which separates Plato from Aristotle: namely the question whether the psyche, being immaterial, is a substance that is separable from the body or not. It is perhaps worth observing that there was no debate in classical times as to whether there *is* something called 'psyche' or not. There is something that makes a difference between animals and rocks, and whatever makes that difference is the psyche. There was some debate about where to draw the line between things that have psyche and things that do not – where do plants fall? – but that there was such a line to be drawn was on the whole agreed. If there was any disagreement, it was from those who held that psyche is to be found in *all* things. Thales is said to have believed this, although it is not altogether clear what he meant by it.[2]

From the earliest times there was a connection between psyche and breath. In the early lyric poets, death is sometimes described as 'breathing out the psyche.'[3] The sixth-century Milesian Anaximenes is quoted as saying that our psyche is air (DK 13B2). That this idea is not in the least incompatible with belief in the survival of the soul after death is evident from the Homeric poems, where the ghosts (the word is *psychai*) of the dead are found in Hades. They are living, or partly living, but they cannot engage in the essentially human activity of communicating until they revitalize their shadowy forms by taking a drink of blood. Among the philosophers the idea of the material soul took on a different aspect. In Heraclitus, the soul is sometimes identified with fire, as in this fragment where the context makes the equation fairly plain: 'For souls it is death to become water, for water it is death to become earth; out of earth water arises, out of water, soul' (fr. 36). Much more is involved in this equation than the material composition of the soul; a little has already been said about it in chapter 4. The important feature is that the human psyche is identified with the symbol of the measured change in the whole cosmos, the 'fire everliving, kindled in measures and in measures going out' (fr. 30). The equation suggests links between the life of a man and the life of the whole cosmos, perhaps also between the intelligence of man and the rationality of the cosmos.[4]

[2] 'And some say the psyche is intermingled in the whole [universe] – hence, presumably, Thales too got the idea that all things are full of gods' (Aristotle, *De anima* 1.5, 411a7).

[3] E.g. Simonides fr. 52; Pindar, *Nemean* 1.47.

[4] For more on the psyche, see Claus, *Toward the Soul*, and his bibliography. There is a wealth of fascinating detail in Onians, *The Origins of European Thought about the Body, the Soul, the World, Time, and Fate.*

A similar link between man and the cosmos is to be found in another line of philosophical and religious speculation. This is the line of thought that connects the human psyche with the stars. There are many strands in the connection: sometimes it is recognizable as the same connection as that between psyche and breath, for the stars are the inhabitants of the upper air or aether; sometimes stress is laid on the quality of the matter that constitutes both psyche and the stars – its purity, freedom from corruption, unchangeability; sometimes it is a question of motion, when the seasonal motions of the heavenly bodies, which give light to the whole earth, are compared to the motions imparted to the body by psyche; sometimes it is simply that the heavenly regions take the place of the underworld as the place to which immortal souls go after the death of the body.[5]

The position of Empedocles with regard to the psyche is fascinatingly ambiguous. There is nothing in his physical system but earth, water, air, and fire, and the two moving forces Love and Strife, and there is evidence that he explained the characteristic activities of the psyche, such as perception and thought, in materialistic terms; moreover, there are periods in the world's history, according to his theory, when all the elements are mixed together in a unified whole, lacking all motion. One would suppose that at such periods all life in the world would be extinguished; yet he can describe himself as 'an immortal god, mortal no longer' (fr. 112.3), and he speaks of 'the *daimones* whose portion is long life' (fr. 115.5). This language belongs to the mystery religions. The soul – but Empedocles does not use the word 'psyche' for this – is thought of as intrinsically pure, but corrupted by association with the body and particularly by eating meat. It is condemned to be punished for its sins by successive incarnations. But an ultimate state of bliss is envisaged in which no further incarnations are required, and the soul is freed from the cycle of birth, death, and rebirth. Nobody has been able to explain satisfactorily how Empedocles could combine this idea with his theory of the four elements and their periodic cycles of mixture and separation. Plato adopted the idea in some of his dialogues; but he was not in the same technical difficulty as Empedocles, because he was in no way committed to a materialist theory of the psyche.

Anaxagoras, the Ionian, shows no traces of allegiance to any doctrine of cyclical death and rebirth, although there is one solitary and very unreliable report that he believed the soul to be indestructible.[6] It is a good deal more interesting that he gave a materialistic description of Mind:

[5] On this subject, see Burkert, *Lore and Science*, pp. 350ff.

[6] Theodoretus V.23 = DK 59A96 (from Aetius IV.7.1, according to Diels, *Doxographi Graeci*, p. 392n.).

It is the finest of all things, and the purest, yes, and it has all knowledge about everything and has the greatest strength. (fr. 12, quoted above, p. 63)

The word translated 'finest' (*leptotaton*) means 'thin,' 'delicate,' or 'fine-textured.' The comparison of Mind with all other things suggests that it is a material stuff, like them. It has been claimed that Anaxagoras clearly intended Mind to be regarded as immaterial, since he repeatedly insisted that it is entirely separate from the mixture of all things: the word *lepton* is not always applied to material things.[7] But this seems to me unjustified. Anaxagoras used spatial terms in describing Mind: it is *in* things (fr. 11), there can be a larger and smaller Mind (fr. 12), and it is *compared* with other substances in texture, as we have said. To say that it is not mixed with anything is to suggest that it is the kind of thing that might be mixed, and is remarkable for not being mixed. We have already seen, in chapter 6, that Anaxagoras gave a reason for supposing that Mind is unmixed: if it were mixed with anything, it would be mixed with everything, and then it could not 'dominate' everything as it does. Its fineness of texture is postulated, no doubt, to explain how it can readily move through all other things in its capacity as the ordering and organizing force of the cosmos.

Aristotle links Anaxagoras with Democritus in this context: both adopted a theory of the material composition of the psyche (Aristotle declares that Anaxagoras in effect identifies Mind with the psyche) in order to give a plausible account of the causation of motion. Democritus, says Aristotle, declared that the soul is warm and fiery, and is made of spherical atoms. The reason for this proposition is that the psyche was thought of as being primarily that which causes motion; causing motion was thought to entail being moved, and the spherical shape was best adapted to permeate the whole of the frame and get it moving.[8] The idea that the atoms of the psyche have some special qualities of shape and size became a traditional part of the atomic theory. The Epicureans adopted it, and worked it out in more detail than Democritus, as we can see from book III of Lucretius' poem (III.177–230). The kind of reasoning that led to the allocation of spherical shape to soul-atoms is the same as that which led to the postulation of hook-shaped atoms to account for the stability of compounds. There is no great gulf fixed between the material of the psyche and that of physical things: the 'laws,' if such an anachronistic term may be permitted, that governed their behavior were similar for both. This, indeed, is a point that draws criticism from Aristotle, who says that Democritus' theory reminds him of the comic poet Philippus' description of Daedalus:

[7] Guthrie, *HGP*, vol. II, pp. 276–7.   [8] *De anima* I.2, 403b28ff.

Daedalus made a wooden statue of Aphrodite – and then gave it the power of movement by pouring quicksilver into it.[9]

Democritus seems to have gone some way toward correlating his theory about the material composition of the psyche with his ideas about thought, emotion, and action.

> Good spirits come to men through temperate enjoyment and a life commensurate. Deficiencies and excesses tend to turn into their opposites and to make large motions in the soul. And such souls as are in large-scale motion are neither in good balance nor in good spirits. (fr. 191, trans. Schofield)[10]

At least this tells us that motions in the soul are directly related to emotional states, and it appears that the preferred state of the soul is one of relative tranquillity of motion. Another much-quoted fragment tends to confirm the general idea:

> Teaching reshapes a man, and by reshaping, re-natures him. (fr. 33)

The Greek word translated 'reshapes' (μεταρυσμοῖ) includes the word used by Democritus to refer to the shape of atoms (ῥυσμός). This is awkward, since the shape of an atom is unalterable. We have to suppose that Democritus used the word here to refer to the positioning of atoms rather than their shape. By repositioning the individual atoms of the psyche teaching reshapes the whole compound. The word φυσιοποιεῖν does not occur elsewhere in extant Greek literature, and is presumably an *ad hoc* invention of Democritus. So I have invented the word 're-natures' to translate it. The meaning is that a man's nature depends on the quality of his psyche, which in turn is a matter of its 'shaping' – that is, the positioning of its component atoms; this nature can be modified by the effects of teaching. A very similar idea is set out at more length in Lucretius' third book (III.294–322).

In Epicurean philosophy, the mortality of the psyche and the impossibility of the survival of an individual personality after death was strongly emphasized. No Epicurean doctrine is better known, thanks to the brilliant and memorable treatment given to it by Lucretius (III.417–623). Of Democritus, there are some dry reports that he believed the soul to be mortal, and there is one hint, at least, that he attacked the myths of the afterlife as being false and ignorant (fr. 297, from Stobaeus). But there is a curious element of doubt about it. He is credited with a book (lost now, of course) entitled *On Those in Hades*. If this is genuine, as it may not be, it is perhaps to be related to a doctrine of his that is mentioned several times in

[9] *De anima* I.3, 406b15ff.
[10] Kirk, Raven, and Schofield, *The Presocratic Philosophers*, 2nd ed., p. 430.

later antiquity, that death may not be a sudden and complete ending of all the phenomena of life. Dead bodies retain for a while traces of heat and even of sensation (DK 68A117, from Alexander and Aetius); hair and nails continue to grow in the tomb (A160, from Tertullian); hence it is hard for doctors to determine when life has ended (A160, from Celsus). Perhaps Democritus wished to offer a rationalistic explanation of certain cases of the revival of those thought to be dead.[11]

Apart from the fragment mentioned in the last paragraph (fr. 297), there is little evidence that Democritus' theory of the nature of man was situated in a polemical context. But there can be no doubt that he took what might be called the humanist view: he believed man to be not a fallen god but a superior animal. In the course of the evolution of living forms in the cosmos, man first emerged 'from water and mud,' or 'from life-bearing moisture.'[12] The Christian Lactantius pointedly contrasts Democritus with the Stoics and Christians:

> The Stoics say it was for the sake of men that the world and all things in it were formed; divine scripture teaches us the same. So Democritus was wrong in supposing that men were emitted from the earth like worms, without maker and for no reason. (*Inst. Div.* VII.7.9)

There is some evidence that Democritus was the originator of the idea that man is a 'little cosmos' or microcosm. There is one direct and explicit attribution of the idea to Democritus by name, in a passage from a certain David, a sixth-century Christian Neoplatonist. In the second century, Galen mentioned that the idea was advanced by 'some ancients who were good on the subject of nature.'[13] The idea is one that through the centuries has played a part in extravagant fantasies, particularly of an astrological kind. But probably not much should be made of it in the case of Democritus, even if the idea is genuinely Democritean. We have already seen plenty of evidence that in the period of philosophy before Socrates the cosmos was often described in terms of an analogy with living creatures. And this is just the obverse of that analogy. If Democritus said that man is a little cosmos, he probably meant that man is made of different parts, each contributing to the ordered working of the whole, just as the cosmos

[11] I am following Guthrie here: see *HGP*, vol. II, pp. 436–8.

[12] Censorinus IV.9 and Aetius V.19.6, both in DK 68A139.

[13] David, *Proleg.* 38.14; Galen, *De usu partium* II.10, K 3 241; both in DK 68B34. It is worth noting that Thrasyllus' arrangement of Democritus' books, listed in Diogenes Laertius IX.45, juxtaposes these two titles: *On Nature, first book. On the Nature of Man (or On Flesh), second book.*

is, and perhaps he meant also that man is like the cosmos in being made of atoms and void.[14]

11.2 *The growth of civilization*

Damp, fire, or bibliophagous worms have made it next to impossible to distinguish Democritus' contribution on this subject. That the fifth century produced rationalizing theories of the origin and development of human society, distinct from the traditional myths, is clear enough, and we have looked at some of the evidence while discussing Anaxagoras. There is enough to prove that Democritus was among those who held such a theory. We are told that he believed that men learnt from spiders how to weave, from swallows how to build, and from swans and nightingales how to sing (fr. 154, from Plutarch); and that music is a fairly recent development (fr. 144, from Philodemus). This shows only that he contributed to this field of speculation, and hardly gives us enough to distinguish his contribution from others. There have been ambitious attempts to discover and assess his theory, but I am not convinced that any of them are successful.[15] For an Atomist's account of the growth of civilization, we must wait for Lucretius.

11.3 *Nature and morality*

The situation is only a little less confused with regard to Democritus' views about the present state of society and the justification of its procedures. In this case the difficulty lies not so much in the lack of direct evidence, because quite a number of surviving fragments say something on this topic, but rather in the lack of a connected argument or context. It is well known that at the time when he lived and wrote, in the second half of the fifth century, there was an extensive debate on the principles of social morality. It would be of great interest to know where Democritus stood in this debate, and it is remarkable that there is so little coherent evidence of his contribution. His fellow-townsman Protagoras was in the forefront, and Plato represented him as conversing with Socrates on this subject, in the dialogue named after him. We might well expect that Democritus too would have much to say about the rival claims of nature and law. But all that remains is a number of disconnected pronouncements – bafflingly difficult to put together into a unified argument.[16]

[14] See also Lloyd, *Polarity and Analogy*, pp. 250ff. and Boas, 'Macrocosm and Microcosm' in *Dictionary of the History of Ideas*, vol. III, pp. 126a–131b.

[15] A brave and scholarly recent attempt, with a good bibliography of others, is Cole, *Democritus*.

[16] For a bold effort to do so, see Havelock, *The Liberal Temper in Greek Politics*.

Something should be said about the debate itself. It happens that one of the longest more-or-less connected fragments of Presocratic philosophy, surviving in two papyri from Oxyrhynchus, deals with the subject. The papyri reproduce parts of a work called *On Truth* by Antiphon, a professional Athenian sophist of the fifth century. They are all concerned with the nature of justice (δικαιοσύνη), and they constitute a philosophical critique of the conception of justice as a virtue that human beings ought to pursue.[17]

Antiphon's argument turns on a distinction between the claims of law on the one hand, and 'what is good for you' (τὸ ξυμφέρον) on the other. 'What is good for you' may be a term borrowed from medical vocabulary; the argument assumes that everyone has an undoubted motive for pursuing what is good for him, not merely for the sake of health, but also for his well-being generally. It is nature itself that determines what is good for you, and one ignores it at one's peril. Justice, on the other hand, is nothing but a matter of conventional agreement, and violations of justice do not automatically lead to punishment. Justice, construed as obedience to the laws and customs of the country, brings no advantages that are grounded in nature; on the contrary, it is nothing but a constraint on nature. 'The things laid down as good for you by the law are fetters on nature,' says Antiphon (fr. 44A).

This case is given eloquent expression by Callicles in Plato's *Gorgias*. Callicles argues for a conception of 'natural justice' that asserts the right of the naturally stronger individual to claim power over others. Nature itself makes one man stronger than another: this is just a fact, and the conventions of society should not be allowed to override it. The passage (*Gorgias* 482c–486d) is too well known to need further description here.

Now although the evidence is fragmentary, it seems clear that Democritus rejected the argument presented by Antiphon and Callicles. He was a supporter of law and custom, and he believed that its benefits are grounded in nature.

> The law wants to be beneficial to the life of men; and it can be, when they themselves want to fare well, for it points out its own virtue to those who obey. (fr. 248)

He explicitly opposed the subversive idea expressed by Antiphon that since the laws are 'fetters on nature' it is best for you to ignore the law when you can do so with impunity:

[17] For a defense of my interpretation, see my article 'Antiphon's Case against Justice,' in Kerferd, *The Sophists and their Legacy*, to be reprinted in my *Cosmic Problems*.

Feel no more shame before men than before yourself; do no more wrong if no one will know than if all men will know. Feel shame before yourself, and establish the law in your soul, to do nothing inappropriate. (fr. 264)

The case for saying that Democritus aimed to ground his morality in his interpretation of nature depends on surprisingly slight evidence: since he certainly wrote extensively on ethical matters, one might have expected more. The clearest link between ethics and physics is to be found in his association of contentment or good spirits (εὐθυμίη) and moderate movements of the soul (see fragment 191, quoted on p. 156). In the context of an atomic account of the nature of the soul, the 'motions' described in this text are presumably to be thought of as movements of atoms.[18] The argument, then, may have run somewhat as follows. The natural constitution of the human soul, made of atoms, is such that disruptive or violent motions are associated with disturbance of mind, and relatively slight motions with tranquillity or contentment. The physical condition of lack of disturbance, and the corresponding calm frame of mind, have an obvious, intuitive claim to be preferred to the opposite condition. So contentment may be accepted as a goal established for men by nature. Diogenes confirms that this was indeed Democritus' designation of the goal, without mentioning its grounding in nature:

Contentment is the goal – not being identical with pleasure, as some have mistakenly understood it, but the quality by which the soul lives in a calm and stable manner, disturbed by no fear or superstition or other emotion. He [Democritus] calls it also 'well-being' (εὐέστω) and many other names. (Diogenes Laertius IX.45)

There is much here and elsewhere in Democritus' ethical fragments that foreshadows Epicurus. We shall return in volume 2 to the debate between Epicurus and the Stoics, who tried to base rival systems of morality on their interpretations of nature.

11.4 *The divinities of the materialists*

In the field of theology we can discern the first clouds in a storm that blew up in a later period and almost drowned the Epicurean Atomists. Democritus had no need of gods in his cosmology: he had explained that each cosmos had its origin in an accidental congregation of atoms in the void, and he believed that the motions that produced a vortex, and from the

[18] See especially von Fritz, *Philosophie und sprachliche Ausdruck bei Demokrit, Platon und Aristoteles*, pp. 32–5; and Vlastos, 'Ethics and Physics in Democritus,' in Furley and Allen, *Presocratic Philosophy*, vol. II, pp. 381–408. I do not think Vlastos' case is seriously undermined by Taylor's criticisms, *Phronesis* 12 (1967).

vortex produced a cosmos, could be explained satisfactorily without agencies such as Mind, or Love and Strife. He asserted that belief in gods was not hallucinatory; but he had left them no part in the creation of the cosmos and the governance of nature. This was a part of the doctrine that led the early Christian Fathers to anathematize the Epicureans.

But Democritus, it seems, was perhaps not a fully fledged Epicurean. He accepted, as Epicurus also did, that the common belief of mankind in the existence of gods must have an objective referent in the material world. Like Epicurus, he posited εἴδωλα, thin, filmy images composed of atoms, to account for it. But unlike Epicurus he appears to have held that these images, in their impact upon men, might do harm or good to them. 'He prayed,' says Sextus (*Math.* IX.19), 'that he might meet with good-luck images.' It is not clear, on the evidence available, whether these images were supposed to be effluents from objectively existing gods, in what I take to be the Epicurean fashion, or self-subsistent entities in their own right.[19]

Socrates was accused of impiety by Meletus and others, and was condemned to death. According to Plato's account of his defense speech, he answered the charge, in part, by showing that he did believe in some divinities, and so could not rightly be called an atheist.[20] The same defense could be put up on behalf of the materialist Presocratic philosophers against the charge of atheism that appears to be aimed at them by Plato in the tenth book of his *Laws*. None of them would have been ready to deny that 'the divine' has some part in the cosmos.

They accepted the existence of divinities, but they had reservations about the traditional gods of Greece. The most famous 'reform' of traditional theology – too famous to require more than a brief reminder here – was the work of the sixth-century Ionian, Xenophanes, who has not hitherto played much of a part in this book because his contribution to the cosmological debate was not of great significance. In theology, however, he had a very distinctive voice:

> Homer and Hesiod, they credited the gods with all the things
> counted as shame and reproach among men –
> thievery, adultery, cheating among themselves. (fr. 11)
>
> The Ethiopians make their gods snub-nosed and black;
> the Thracians make them grey-eyed and red-haired. (fr. 16)
>
> If oxen, horses, or lions had hands,

[19] For further discussion of Democritus' theology, see Guthrie, *HGP*, vol. II, pp. 478–83. Some disagree about Epicurus' position: see David Sedley, in Long and Sedley, *The Hellenistic Philosophers*, vol. I.

[20] *Apology* 27c–e.

> or could draw with hands or accomplish the arts of men,
> horses would copy horses and oxen would copy oxen
> when they drew the forms of gods, and would make their bodies
> just like the shape they each had themselves. (fr. 15)

The point of this sardonic wit was not to banish gods from the scene, but to substitute a more sophisticated theology for the simple anthropomorphism of the myths:

> *One* god, greatest in the company of gods and men,
> like mortal men neither in his shape nor in his thought. (fr. 23)
>
> Without toil by the thinking of his mind he shakes the universe. (fr. 25)

One of the primary characteristics of a god is his immortality. It was perhaps only to be expected that divinity would be attributed by the cosmologists to those beings in their theories that were regarded as eternal. The Boundless in Anaximander's system (to go right back to the beginning) was said to be 'divine,' according to Aristotle, because it was 'immortal and indestructible.'[21] Most strikingly, Empedocles gave the names of divinities to what he otherwise called the four 'roots,' earth, water, air, and fire, and the reason is no doubt the same: they are things that 'live for ever.' He also called the motive agencies of his system by the names of gods: Kypris or Aphrodite (frs. 86, 87, 98) and perhaps Deris (fr. 122). Werner Jaeger found sufficient material in the Presocratic philosophers before the Atomists for a series of Gifford lectures, published as *The Theology of the Early Greek Philosophers*.

Jaeger did not hesitate to treat Anaxagoras' conception of Mind as a contribution to theology. This was bold, because the fragments do not directly call Mind by the name of god or the divine. The evidence he relied on was the hymn-like style of fragment 12, which we have already noticed (above, p. 63), and the epithets attached to Mind in that fragment: 'infinite,' 'self-ruling,' 'itself by itself.' This may be thought insufficient to bear the rather heavy weight Jaeger put on it. A similar idea is also attested for the successor of Anaxagoras, Diogenes of Apollonia, whom we looked at briefly in chapter 6. Diogenes believed himself to have located Mind in the air that permeates the whole cosmos and is thereby in a better position than anything else to 'steer' and 'control' everything:

> It seems to me that what possesses intelligence is air, as men call it, and that by this all are steered and it controls all. For this itself seems to me to be god, and to have penetrated everywhere, to dispose all things, and to be present in everything. (fr. 5)

This would be conclusive evidence for a kind of theism, if only the text

[21] Aristotle, *Physics* III.4, 203b11 = DK 12A15.

were certain. The word 'god' (*theos*), however, is an emendation: the manuscripts have the untranslatable 'custom' (*ethos*). The emendation is confirmed by Theophrastus' statement of Diogenes' theory: 'the inner air is perceptive, being a small portion of god.'[22] And most editors now accept it.[23] It is also confirmed in a general way by Aristophanes' *Clouds* – for the Clouds are airy *divinities*, and there is little doubt that Aristophanes was making fun of Diogenes' theories.

Aristophanes at the same time gives us a clearer idea of what was involved in the charge of atheism in Athens in the late fifth century. The elderly countryman Strepsiades is shocked by the statements of 'Socrates' that rain has a materialistic explanation and is not caused, as he had thought, by Zeus urinating through a sieve (*sic*! – he *is* a countryman), and that it was the 'aerial vortex' that brought the world into being, rather than Zeus again. Anaxagoras, who was at some time charged with impiety and exiled from Athens, offered an explanation of lunar eclipses that was substantially correct, and was in trouble because of this, 'since they did not tolerate the natural philosophers (φυσικοί) and sky-babblers (μετεωρολέσχαι) as they were called, on the ground that they frittered away the Divine into irrational causes, non-providential powers, and necessitated accidents.'[24] He was notorious for the doctrine that the sun is an 'inflamed rock.'[25]

It appears that popular opinion averted its gaze from the reformed theology adopted by the *physikoi*. So far as the evidence shows, they were not found to have violated contemporary customs in ritual, either in public rites presented by the customs of the state or in the everyday practice of rituals in the home. What would provoke the wrath of the populace, however, especially if it was stimulated by suitable rhetoric, was a theory that reduced the traditional sky-gods and their actions to nothing but matter in motion. The populace sometimes refused to give up Zeus and his thunderbolts in favour of clashes of the hot and the cold, in much the same frame of mind as those who today cling to a literal interpretation of Genesis 1.1–2, and think that the Darwinian theory of evolution is a form of wickedness.

11.5 *The progress of rationalism*

We began this history of early Greek cosmology, in chapter 3, with a myth

22 *De sensibus* 42.
23 See most recently Laks, *Diogène d' Apollonie*, pp. 49–50.
24 Plutarch, *Life of Nicias* 23.
25 This is mentioned by Xenophon, Diogenes Laertius, Harpocration, the Suda, Olympiodorus, and others (DK 59A1, 2, 3, 19, 20a, 73).

of the cosmos in which the agents were anthropomorphic gods and the events were sexual acts, births, and the results of human intentions. We interrupt the story here, at a moment when the human analogies have been minimized, although not wholly eliminated, and cosmology has been reduced, so far as possible, to matter-in-motion – inert matter in automatic motion. This is the time to pause briefly to take stock of the material we have been looking at. Two questions in particular may be asked: what is the general nature of the mental processes that created this development, and what are its external causes, if any can be identified? Both questions have been extensively examined many times, and I shall not attempt to do more than pick out a few themes for special emphasis, acknowledging gratefully the work of others.[26]

Remarkably little of the development we have been tracing was provoked – or even aided – by empirical observation. It is true, of course, that astronomical observation provided the foundation for much of Greek cosmology. But even in this area, a surprisingly large proportion of the observational records were apparently inherited by the Greeks from the earlier Near Eastern civilizations of Egypt and Babylon. For instance, it was not any new observation that led to the discovery, in Greece, of the explanation of lunar eclipses, but rather a new way of thinking about what had long been observed.

Concerning the nature of our earthly environment, a few observations were of course of basic importance: the vertical fall of heavy objects and the rise of light ones; the vaporization of water when heated, and its solidification when frozen; the regular phenomenon of the changing seasons, and exceptional and unpredictable phenomena such as gales, lightning, and earthquakes. Several more could be added to the list. But not much of the achievements of early Greek cosmology can be attributed to controlled observation, still less to experiment.

It is relevant to observe that at the same time as Democritus was developing the atomic theory, the study of nature was being carried on and expounded in prose treatises of a different kind, by Hippocrates of Cos and other medical writers whose work was subsumed under his name.

[26] Especially Lloyd, *Magic, Reason, and Experience*. There is very little that I would dissent from in this book, and little that I would want to add; it is also invaluable for having an extensive and recent (1979) bibliography of the subject.

There are three other items to which I would give special emphasis. Cornford's *Principium Sapientiae* (1952) was one of the books that started me on this subject, although there is much in it that I would disagree with now. Popper, *Proceedings of the Aristotelian Society* 59 (1958–9), makes one of the most important points. Vernant's *Les Origines de la pensée grecque* (1962) was very little noticed by English-speaking scholars, but perhaps the English translation (1982) will remedy that.

Sometimes their subject matter overlapped with that of the cosmologists, but their intention was of course different. The Hippocratic essay called *Airs, Waters, Places* begins by stating that the student of medicine must know about the seasons of the year, the character of the winds that blow in his region, the nature of the water supply, and the quality of the local soil. The author explicitly notices that those may seem to be subjects for speculation rather than practical medicine, but he claims that such knowledge does in fact have practical uses:

For knowing the changes of the seasons, and the risings and settings of the stars, with the circumstances of each of these phenomena, he will know beforehand the nature of the year that is coming. Through these considerations and by learning the times beforehand, he will have full knowledge of each particular case, will succeed best in securing health, and will achieve the greatest triumphs in the practice of his art. If it be thought that all this belongs to meteorology, he will find out, on second thoughts, that the contribution of astronomy to medicine is not a very small one but a very great one indeed. For with the seasons men's diseases, like their digestive organs, suffer change. (*Airs, Waters, Places* 2, trans. W. H. S. Jones)

It is in the realm of medicine, above all, that an empirical approach to the natural world might be expected. And in fact the Hippocratic books are full of the records of observation of the progress and treatment of illness; there is even some evidence for the practice of dissection (of animals) for the purpose of research.[27] But when the doctors turn to framing theories about the nature of man, and ponder the causes of disease and cure, their methods are not so very different from those of the cosmologists. It is true that there are attacks in the Hippocratic books on the speculative habits of the natural philosophers. One much-quoted example comes from the treatise called *On Ancient Medicine* (less misleadingly translated *Tradition in Medicine* in the collection from which I quote).[28] It is an attack on those who base physiology on the primary 'opposites' of physics:

I am utterly at a loss [says the writer] to know how those who prefer these hypothetical arguments and reduce the science to a simple matter of 'postulates' (*hypotheseis*) ever cure anyone on the basis of these assumptions. I do not think that they have ever discovered anything that is purely 'hot' or 'cold,' 'dry' or 'wet,' without it sharing other qualities . . . It would be useless to bid a sick man to 'take something hot.' He would immediately ask 'What?' (ch. 15, trans. J. Chadwick and W. N. Mann)

[27] See Lloyd, *Magic, Reason, and Experience*, pp. 156ff., *Sudhoffs Archiv* 59 (1975).
[28] Lloyd (ed.), *Hippocratic Writings*.

Nevertheless what the writer proposes to substitute for these theoretical 'postulates' is not much more empirically based or checkable than what he attacks. He merely substitutes a list of more complex items, such as the bitter-and-hot, the salty-and-hot, and so on.[29]

One important quality must be recognized in the writings of these early Greek thinkers, both cosmological and medical. It is evident that they were not written down in the same spirit in which a law is written down, as an authoritative pronouncement that is to be accepted and obeyed. They were offered for criticism; at any rate, they received criticism. If anything has emerged clearly from the earlier chapters of this book, it is that the tradition grew from one generation to another through criticism of the theory and substitution of another thought to be better than the first.[30]

But what in general terms can be said about the criteria that were used to establish that one theory was better than another? In the first place, we can follow the guidance of Parmenides, who is one of the few in the early period whose methodological remarks survive. 'Judge by reason (*logos*) the hard-hitting refutation (*elenchos*) that I have uttered,' says the goddess in his poem (fr. 7.5). *Elenchos* may refer to a process of examining a person, as in a law-court, to determine if he is telling the truth, or to the testing of a proposition by attempting to refute it.[31] The grounds on which something is regarded as refuted may vary, of course. In the case of Parmenides, logical consistency was the test; in the case of a lying sea-captain in a story in Herodotus, to quote a non-philosophical example of *elenchos*, his story of the death of a passenger was refuted by the actual appearance of the passenger alive (1.24). But the general character of the methods of the cosmologists is summed up well in Parmenides' words 'judge by reason.' What was to be accepted – the *elenchos* itself or what survived the *elenchos* – was what appeared reasonable. If the test was not one of logical consistency, it was frequently a matter of appeal to analogy, often an analogy between the cosmos and living creatures, as we have seen. From the beginning, too, we can observe the powerful grip of the concept of simple opposition in cosmology: theories involving the hot and

[29] See further two papers reprinted in Furley and Allen, *Presocratic Philosophy*, vol. 1: Cornford, 'Was the Ionian Philosophy Scientific?' and Vlastos' review of Cornford's *Principium Sapientiae*. See also Miller, *Transactions and Proceedings of the American Philological Association* 83 (1952) and 86 (1955).

[30] Did the Pythagoreans stand outside this critical tradition? There is ample evidence that rules of secrecy were imposed on members of the Pythagorean community, as in mystery cults, and it seems that cosmological doctrines were not exempt from these rules. See Guthrie, *HGP*, vol. 1, pp. 150–3.

[31] See Lesher, *Oxford Studies in Ancient Philosophy* 2 (1984), and my article 'Truth as what is unrefuted,' in my *Cosmic Problems*.

the cold, the dry and the wet, and other pairs of opposites persist through the Presocratic period as far as Aristotle. This may well have been an inheritance from the conceptual world of cosmogonic myth, where the primary oppositions are between male and female, and god and man.

The tests of logical consistency, analogy with something more familiar, and degree of fit with an established conceptual framework apply, of course, to fields of argument far wider than cosmology alone. They are familiar elements in all kinds of debate, and especially in arguments in courts of law and in political assemblies. It is in this area, if anywhere, that we may find some elements for a causal explanation of the rise of cosmological theory in Greece in the sixth and fifth centuries.

It has been suggested that we should look for such an explanation in contacts between philosophical speculation and newly emerging technologies. But the evidence is unconvincing. Even less plausible is the suggestion that different philosophical theories are reflections of different forms of social organization – that Empedocles' theory of Love and Strife, for instance, is somehow connected with the antagonism between freeman and slave, or that atomism is an ideological reflection of individualism in society.[32] It may be that the whole idea of a causal explanation is futile: perhaps the whole intellectual movement is best regarded as the work of a few untypical and inexplicable individuals. But it is hard to resist the suggestion that the new structures of the Greek *polis* gave a new importance to the arts of speech (*logos*), and that the new practices of *logos* stimulated the desire and the ability to frame a *logos* about the cosmos.

It was in democracies that the power of *logos* was most developed and most valued, and in democratic Athens, as everyone knows, that it reached its fullest development. The Ionian cities like Miletus, where the first cosmologists lived, were not democratic. But they were politically independent, they were not monarchic (except for occasional periods of tyranny), and they were in a state of frequent political change. Political argument was both possible and significant: there were people to be persuaded.[33] Political argument is mainly tied to its own time and place, and none of it has survived from the time before Herodotus and Thucydides and the early Attic orators. But we need not doubt its existence. From the middle of the fifth century, the sophists made a living largely by teaching young men in many cities of Greece to be persuasive public speakers, and this was a market that did not spring into existence in a few years. We may

[32] Thomson, *Studies in Ancient Greek Society: the First Philosophers*, pp. 308–14.
[33] See Emlyn-Jones, *The Ionians and Hellenism*, especially chs. 5 and 6.

reasonably assume that the habit of persuasive argument in a public context was one of the main reasons why philosophical cosmology flourished in Greece in the sixth and fifth centuries, and not in other places at other times.

The growth of literacy in Greece must be mentioned also as part of the background to this history, but its role is less clear. This was certainly the time when Greek culture was in the process of transformation from one in which the transmission was predominantly oral to a literate one.[34] But other cultures became literate, and did not develop philosophical argument. It is easier to invoke the change from an oral to a literate culture in explanation of the particular forms adopted for philosophical discourse than to show that it played a causal role in the emergence of philosophy as a distinct literary genre.

[34] See especially the writings of Havelock: *Preface to Plato* and *The Literate Revolution in Greece and its Cultural Consequences*. The latter has a most illuminating discussion of Heraclitus and Parmenides, viewed within the context of an oral culture (pp. 240–56). There is a set of essays on Havelock's thesis edited by Robb: *Language and Thought in Early Greek Philosophy*. See also the review of this by Ferrari, *Ancient Philosophy* 4 (1984).

12 Plato's criticisms of the materialists

In chapter 2 we have already taken a preliminary look at Plato's first observations on the cosmology of his predecessors. He gave us an unforgettable picture of Socrates reviewing his life of philosophizing, as he sat in prison with friends on the last day of his life. The main target of his criticism of earlier philosophy was its neglect of the question 'to what end?' or 'what is the good of this?' The physicists gave him nothing but a story of matter in motion, of how, not why, things come to be as they are. This Socrates found so unsatisfying that he turned to a quite different kind of philosophy.

Plato returned to criticism of materialism in works written later than *Phaedo*, and that is the subject of this chapter. First we must observe that Plato seldom names the targets of his criticism. Of the 'physicists' we have reviewed in earlier chapters, Plato mentions Thales, but not Anaximander or Anaximenes; Anaxagoras several times, but Empedocles only twice; and Leucippus and Democritus not at all. The omission of the last two has surprised the commentators from ancient times to the present day.[1] There is a story in the biography of Democritus in Diogenes Laertius (IX.40) that Plato wanted to burn all of Democritus' books that he could collect, but was prevented by two Pythagoreans. It was no use, they said, the books were too widely published. In the same source (IX.36) Democritus himself is quoted as saying 'I came to Athens and no one knew me.' These stories may be just graphic ways of illustrating what we can learn from the text of Plato: that Plato was radically opposed to the philosophy of Democritus, and that he never mentioned him directly.

This silence means that we must look below the surface for Plato's reaction to the Atomists. Some parts of his own cosmology can with some plausibility be interpreted as positive developments of Atomism – doctrines held by the Atomists that were modified by Plato and incorporated into his own world system. These will be discussed in volume 2. Our subject at present is not what Plato accepted, but what he rejected, and why.

12.1 *Matter as a secondary cause*

Plato returns to the theme of our *Phaedo* passage in the *Timaeus*. After an

[1] Diogenes Laertius III.25: 'It is asked why he did not mention Democritus.'

account of the physical mechanism of vision, involving the texture of the eyeball and different kinds of light-producing fire inside and outside the body, he continues:

> Now all these things are among the accessory causes that the god uses as subservient in achieving the best result that is possible. But the great mass of mankind regard them, not as accessories, but as the sole causes of all things, producing effects by cooling or heating, compacting or rarefying, and all such processes. But such things are incapable of any plan or intelligence for any purpose. For we must declare that the only existing thing that properly possesses intelligence is soul, and this is an invisible thing, whereas fire, water, earth, and air are all visible bodies. And a lover of intelligence and knowledge must necessarily seek first for the cause that belongs to intelligent nature, and only in the second place for that which belongs to things that are moved by others and of necessity set yet others in motion. We too, then, must proceed on this principle: we must speak of both kinds of cause, but distinguish causes that work with intelligence to produce what is good and desirable, from those which, being destitute of reason, produce their sundry effects at random and without order. (Plato, *Timaeus* 46c–e, trans. Cornford, adapted)

This is barely more than a dogmatic repetition of the point made earlier in the *Phaedo*. We need to recognize in the world what is 'good and desirable,' as opposed to things that are 'random and without order.' And we are to accept that only intelligence, which is the property of an immaterial thing, the soul, can produce good and orderly effects: matter in motion, although it contributes to the explanation of the cosmos, can never provide the whole explanation, since it is without mind. Plato skillfully introduces this point, which might have found a place virtually anywhere in his discourse, immediately after his account of sight. Sight is one of the most self-evidently suitable aspects of the natural world to give force to his argument. Clearly it cannot occur without the right material elements. Yet it is hard to believe that the material elements alone will explain the facts of sight. We know well enough how one piece of matter can transmit a color to another, as in a process of dyeing; but that one *sees* a color – that is a very different thing. It is not just that one thinks the eye is too intricate a device to be the product of undesigned motions. There is also the feeling that something of a quite different order is involved, wholly beyond what can be explained by varieties of fire and reflecting surfaces.

So Plato makes his point again, that we cannot neglect matter-in-motion in our explanations of the physical world, but we must always regard it as a secondary factor. Matter-in-motion is just what is *necessary*, it does not of itself produce the good:

All these things, then, being so constituted of necessity, were taken over by the maker of the fairest and best of all things that become, when he gave birth to the self-sufficing and most perfect god [*sc.* the cosmos]; he made use of causes of this order as subservient, while he himself contrived the good in all things that come to be. We must accordingly distinguish two kinds of cause, the necessary and the divine. The divine we should search out in all things for the sake of a life of such happiness as our nature admits; the necessary for the sake of the divine, reflecting that apart from the necessary those other objects of our serious study cannot by themselves be perceived or communicated, nor can we in any other way have part or lot in them. (Plato, *Timaeus* 68e–69a, trans. Cornford)

12.2 *Matter and predication*

In the *Sophist*, Plato's subject is the question of what things there are in the world, and at one point he imagines an encounter with people 'who pull everything down to earth from heaven and the unseen, literally clutching rocks and oaktrees with their hands.' They declare that there is only that which offers resistance and contact: they identify being with body (246a). The interlocutor in Plato's dialogue concedes that he has met many such people – so we are dealing with the popular version of philosophical theories, rather than with the words of some particular philosopher. Plato's simple purpose, in the context, is to force such people to relent a little from their metaphysical austerity. They must concede that there are living beings, that living beings have a soul as well as a body, and that the soul has properties like justice and wisdom – and these, which they have now included in the list of what things there are, are things that cannot be seen or touched.

If this line of criticism were directed at Leucippus and Democritus, it would not hold them up for long. Their ontology includes a vast range of secondary properties, like colors or textures, which belong to compounds by virtue of the shape, size, and arrangement of the component atoms. Items like justice or wisdom, in their theory, are simply the kind of property produced by particular arrangements of soul atoms. In accordance with fragment 125 of Democritus (quoted above, p. 133), they have to concede that such items are 'by custom,' rather than 'in truth' (the latter description is reserved for atoms and void). But they felt no difficulty in giving some kind of status, as 'things that are,' to such secondary properties.

Those who adopted the material theory of Anaxagoras might, however, be more troubled if Plato pressed this point against them. If a thing is white or blue or black, according to them, it is because there is a 'predominant' amount of white or blue or black in the material ingredients that compose it. So far as we can tell from the fragments, Anaxagoras held that each predicate that picks out an item among the things that there are in the world names a material ingredient in the original mixture. There is Mind, of course, standing apart from the universal mixture, but nothing else. So it would appear that Anaxagoras must either claim that justice and wisdom are material ingredients, or else concede that they are not among the things that really are – or perhaps treat them in some way as properties of Mind. How he handled this problem is not known.

It is my belief that reflection on this difficulty in Anaxagorean physics helped to steer Plato towards the theory of Forms. Plato dealt with the problems of predication with the help of this theory, and it is instructive to compare him with Anaxagoras.[2] To take an example, in the predicative statement 'this book is blue,' we ask what exactly are the items referred to by the subject phrase and the predicate, and what is the relation between them. Anaxagoras' answer would be that the subject phrase picks out a particular area of the universal mixture, and the predicate names a kind of stuff, some of which constitutes a predominant material ingredient of this area. The relation between them is just that the item named by the predicate is an ingredient – and a predominant one – in the item named by the subject phrase. Plato's objection to this as a general account of predication is that there are many predicates that do not seem to name ingredients drawn from a universal store of material: 'large,' 'equal,' 'just,' 'wise' would be examples. But he wanted to retain as much as possible of the Anaxagorean theory. What he retained was Anaxagoras' idea of the relation between the item referred to by the subject phrase and the referent of the predicate: the subject 'has a share of' or 'participates in' the predicate. Since it did not seem to him to make sense to think of the predicated items as masses of material substance, as Anaxagoras did, he postulated immaterial beings instead, called by the general name 'Form' (εἶδος) or 'Idea' (ἰδέα).

But this left a set of problems in the notion of participation. Plato tried to make them less troublesome by insisting that his notion of participation is not a materialistic one. But it is not clear that he ever regarded them as solved. In the *Parmenides* (131a–c) he puts the problem in the form of a dilemma: either the whole of the predicate item is present in the subject, or part of it is; and both options have objectionable consequences. In the

[2] Rosenmeyer's early paper, *Transactions and Proceedings of the American Philological Association* 88 (1957), is relevant here, although he does not make the comparison with Anaxagoras. The comparison between Anaxagoras and Plato is further explored in my article 'Anaxagoras' (cited on p. 61 n.1). The idea that the theory of Forms owes something to Anaxagoras is speculative, but gains some support from the structure of Socrates' autobiographical speech in *Phaedo* 97c–102a. See also Brentlinger, *Phronesis* (1972).

Philebus (15a–b), he presents the same dilemma: either the Form is one thing, in which case it cannot be 'shared' by many particulars, or it is shared, and in that case it has no unity. Such problems show that we cannot think of Platonic Forms as being of exactly the same kind as Anaxagorean 'things' (χρήματα). But it was evidently easier to say what Forms are not than to say what they are, and Plato's reticence has left an enduring problem to his commentators.

12.3 *The impiety of materialism*

In his last dialogue, the *Laws*, Plato constructs a legal code for a model Cretan city. In the tenth book he reaches the subject of offenses against the gods of the state. He distinguishes three wrong views about the gods that are sometimes held: that they do not exist, that they exist but take no thought for the human race, and that in the event of offenses against them they can easily be placated by prayer and sacrifice. The first category of mistake is exemplified by people who believe the cosmos to be a collection of matter-in-motion, and who explain the motions of the sun, moon, and stars, and the seasons of the earth without positing any divine agency. Plato imagines these people as demanding persuasion, rather than threats and punishment, before they will recant. So he constructs a rational criticism of their materialistic atheism.

Before examining this criticism, we must pause to ask some questions about its target. As usual Plato mentions no names, and he gives few clear hints. If we study the doctrines of his predecessors we find it hard to identify any particular philosopher or school that satisfies all of Plato's description in precise detail. It is not clear that any of the Presocratic philosophers were atheists.[3] Several held that the heavenly bodies are material objects rather than gods, and Anaxagoras is known to have been prosecuted for impiety in Athens on this ground. But the account of motion criticized here is nothing like Anaxagoras' theory of a cosmic Mind. Perhaps Plato had the Atomists in view; but there is nothing that makes us think of known details of the atomic theory. Trevor Saunders, in his translation of the *Laws*, writes: 'Plato is, in my view, attacking a "climate of opinion," or an amalgam of several views, rather than a single, clear-cut doctrine.'[4]

What, then, is the content of this amalgam, to what extent is it derived from the theories of the cosmologists, what did Plato find objectionable

[3] See chapter 11 for more about the theological beliefs of the early philosophers.
[4] *Plato: the Laws*, p. 409. For an analysis of the text, see Martin, *Studia Philosophica* 11 (1951), and Morrow, *Plato's Cretan City*, pp. 470–96.

about it, and how did he propose to 'persuade' its supporters to change their minds?

Atheism, says the Cretan Cleinias, is easy to refute: you have only to point to the wonderful order of the cosmos and the regular march of the seasons. This might serve, the Athenian replies, to refute those who are just attracted by the freedom of the godless life, but it will not convince the much more dangerous type of atheist, no doubt unfamiliar in Crete, who suffers from 'a form of ignorance ... which passes for the height of wisdom' (886b). In Athens, there are books: not only the harmless ancient myths about the births of the gods, but also those which teach that the sun, moon, stars, and earth – the Cretan's prime evidences of the gods' activity – are all of them nothing but rocks. The theory at the back of this doctrine is that everything is the product of nature, art, or chance (888e), and that nature and chance are prior to art. It is worth quoting a little:

> Fire, water, earth, and air – all these, they say, are by nature and chance, none of them by art. And the bodies that are secondary to these – meaning earth and sun and moon and stars – have come into being through these entirely soulless agencies. Each of these moves by chance according to the power that belongs to it. However it happened that they fitted together in some appropriate way – hot with cold, or dry with wet, soft with hard, everything that blended together by chance and from necessity in this blending of opposites – in that way and just by virtue of these things they generated the whole heaven and everything in it, and animals and all the plants, too, since the four seasons came from these. And it wasn't because of Mind, they say, nor because of some god nor by art, but just – as we keep saying – by nature and chance. (*Laws* x, 889 b–c)

The Athenian goes on to accuse these atheists of a moral heresy, as well as a cosmological one: they declare that both the gods, and the standards of morality upheld by the gods, are matters of artifice not grounded at all in nature, and they uphold 'the right way of life according to nature, which is nothing but a life of mastery over others and not being a servant to the rest as the law enjoins' (890a). Here Plato manages to bring together the cosmologists of whom he disapproves and the sophists like Antiphon, who had advanced arguments in favor of natural advantage over the claims of the laws and customs of society, as we saw in chapter 11.

The Athenian's reply to atheism depends on one single point: a question of the cause of motion. He offers a number of distinctions between different kinds of motion, ending with that between the kind of motion that can bring about other motions but not itself, and the kind that can generate both itself and other motions. The latter is agreed to be prior to all other motion. 'If, as most of these people dare to say, all things should come together at a standstill' – a clear reference to Anaxagoras' theory of the beginning of the cosmos, 'all things together' – then what must be the first motion thereafter, of those that have been distinguished? Clearly, self-generating motion. But self-generating motion is found only in living things, and soul is what is always associated with life. Hence soul is prior to body or matter, and it is soul that is responsible for the movements of everything in heaven, earth, and sea. The motions of the soul, such as wish, reflection, care, decision, judgement true and false, joy and grief, confidence and fear, hate and love, 'take over the secondary motions of material bodies and guide all things to growth and decay, separation and composition, and the accompaniments of these motions, such as heat and cold, heaviness and lightness, hard and soft, white and black, bitter and sweet' (897a).

There is obviously little difference between this argument in the *Laws* and the criticisms I have already quoted from the *Phaedo* and *Timaeus*: the motions of matter are secondary to the motions brought about by soul or mind. But there is some difference: it is only in the *Laws* that we find the distinction between self-generated motions and the rest, and the emphasis on soul as the locus of self-generated motion.

This point has little force against Anaxagoras and Empedocles. Both were ready to concede to Plato that moving agents must be added to the list of material components of physical bodies. There might indeed be some argument between them as to whether Mind, 'the most finely textured of all things,' or Love and Strife, 'equal in length and breadth,' could be as capable of self-motion as Plato's wholly immaterial soul.[5] But if the dispute is simply as to whether nature and chance are prior to art or not, neither Anaxagoras nor Empedocles would have found it too difficult to agree with Plato that they are not. Both might be uneasy about the term 'art,' but they could agree at least on a mover or movers in some sense transcending physical matter and exercising some sort of qualitative control over the motions of matter.

With Leucippus and Democritus, Plato's dispute would be a different matter. They did indeed hold that the motions of material bodies are primary, and that the soul and its motions (which must include art) are derivatives from these motions. But they would answer Plato with the charge that he was begging the question: it was still to be decided whether the phenomena of life, including self-motion, could be plausibly explained as derivatives of inanimate elements. Plato had no right to assume, without argument, that they could not. As to Plato's further point that if the universe has a motionless phase self-motion must be the first to emerge sub-

[5] For the significance of these descriptions, see above, sections 6.2 and 7.2.

sequently, the Atomists need not be troubled by it, because they did not believe in a motionless phase: eternal motion was the condition of the atoms. Plato's point is a chronological one: adopting the premisses of the Anaxagoreans for dialectical purposes, he demands an account of the beginning of motion in time. The Atomists simply denied that there was any such beginning. There is a more profound point, raised by Aristotle, as to whether there must be an a-temporal *principle* of motion transcending all of the everlasting motions of the physical atoms. This we shall consider in the next chapter.

To return briefly to the *Laws* of Plato: he now distinguishes between motion that is regular and orderly, and disorderly motion. The heavens exhibit regular motions: they must be directed by the best kind of soul – that means, by gods. The case against the atheists is complete.

As we shall see in the second volume of this book, the regularity of the motions of the heavens was one of the biggest stumbling-blocks in the way of the progress of Atomism. The other major obstacle was the problem of the generation of living species from inanimate matter. Plato succeeded, in this section of the *Laws*, in setting a good deal of the agenda for the next few centuries of debate.

Plato's proposed legislation to remove atheists from his Cretan city is quite remarkably stern and oppressive. For those not persuaded by his case for theism he proposes penalties that range from five years in prison to solitary confinement for life. 'In this ominous passage [907d ff.] Plato becomes the first political thinker to propose that errors of opinion be made crimes punishable by law.'[6] The proposed statute is 'without parallel in any surviving code of ancient Greece.'[7]

[6] Morrow, *Plato's Cretan City*, p. 488.   [7] Vlastos, *Plato's Universe*, p. 23.

13 Aristotle's criticisms of the materialists

Aristotle vigorously attacked the Atomists and their predecessors for their theories of the elementary material components of the physical world. He himself held a theory that there are such elements, but it differed from all that had come before: he does not even spare Empedocles, who anticipated him at least in picking out earth, water, air, and fire as the elements. His criticisms are directed at the character given to the elements, and the use made of these elements in explaining the phenomena.

We shall examine these criticisms of early theories of the material elements in the second section. First, we must look at a more fundamental criticism – a criticism of materialism itself. Aristotle argues that an account of the material elements is a necessary part of explanation in the study of the natural world, but it is not sufficient for explanation, except in a minority of less important cases. He accuses his philosophical predecessors of a crucial failure, in that they virtually ignored form and finality in nature. This takes us right to the heart of the differences between Atomism and Aristotelianism.

13.1 *The failure of Presocratic explanations*

The subject under discussion is nature, *physis*. At the beginning of the second book of his *Physics*, Aristotle makes some very general remarks, which then form the basis for his criticisms of the Presocratics. Things that are 'by nature,' he says, are the animals and their parts, the plants (and their parts, too, presumably), and the simple bodies (earth, water, air, and fire); and the characteristic that distinguishes them from things that are not by nature is that they have an *internal* principle of motion and rest peculiar to each kind. Thus a stone *falls* toward the center of the world, if it is free to do so, plants *grow*, animals *seek food*. Artificial things have no such motion that belongs to them just because of what they are; if they have an internal principle of motion, it is because they are made of natural materials. A stone statue falls because it is stone, not because it is a statue. This is a criterion that fails to work in the machine age, but in a donkey technology it worked well enough.

Where, then, is this internal principle to be located in the things that are by nature? Either in the matter or in the form. One line of thought, Aristotle says (II.1, 193a10ff.), led some philosophers to locate nature in matter: if you were to plant in the ground a wooden artefact, like a bed, said Antiphon the Sophist, what would grow, if anything could, would be not a bed, but wood – the material of which the bed is made. So the wood, it seems, must be the bed's nature. If that material is itself to be regarded as a modification of some more basic material, then the latter must be identified as the nature of the things that exist – water, or fire, or all four of the elements.

But the argument from the planted bed can be used in a quite different way, Aristotle continues. The bed is a product of art, not of nature. The carpenter's art gives the bed its characteristic form: the art is to be located in the form, not in the rough-sawn lumber of which the bed is made. But man is a product of nature. Beds are not begotten by parental beds, but men are begotten by men. So if the form of the bed is the locus of the relevant *art*, and if nature is the analogue of art in the case of a man, it follows that the *nature* of a man is to be located in his form rather than in his matter. And this is true of all things that are 'by nature.'

Aristotle returns to this point in other contexts. One example, and a very significant one, is in the closing chapter of *Metaphysics* VII. When we are considering objects that exist by nature, we are considering organized wholes: they are not like heaps, he says, but like syllables:

> BA is not the same as B and A, nor is flesh ⟨just⟩ fire and earth ... the syllable, then, is something – not only its elements (the vowel and the consonant) but something else ... ⟨This something else⟩ is the cause that makes *this* thing flesh and *that* a syllable. And similarly in all other cases. And this is the substance of each thing (for this is the primary cause of its being); and since, while some things are not substances, as many as *are* substances are formed in accordance with a nature of their own and by a process of nature, their substance would seem to be this kind of 'nature,' which is not an element but a principle. An element, on the other hand, is that into which a thing is divided and which is present in it as matter; e.g. A and B are the elements of the syllable. (*Metaphysics* VII.17, 1042b12–34, trans. Ross)

This introduces some of the terms of Aristotle's own metaphysical theory, which cannot be adequately explained in a few words, but his thought can be roughly paraphrased as follows. For any object (he thinks of medium-sized objects accessible to sense-perception) we can distinguish between that constituent or set of constituents of it that makes it what it is, and all other constituents that it may have; we call the former its *substance* (*ousia*, 'being'). Thus there are some characteristics of the tree outside my window that make me sure it is a willow, and they do not include its being 45 feet tall, in the north corner of the garden, outside my window, and so on. Where are we to find the substance of a thing? What sort of constituents go to make up the substance? Notoriously, Aristotle does not always answer these questions in the same way, but in the passage just quoted he asserts that it is the 'something else' that makes B and A into the syllable BA, or a certain quantity of the simple bodies into flesh or bone. What he has in mind is the form of the thing; it is that which is expressed in the definition; sometimes he calls it the *logos*.

Exactly how we are to think of the 'being' of this substance is a difficult question. For the present it is enough to notice that Aristotle distinguishes it firmly from the material elements of which a thing is composed. 'An element is that into which a thing is divided': when it is so divided, the elements persist but the substance is destroyed. So the substance is not identical with the elements, nor reducible to the elements: it is 'something else.'

There is no direct criticism of the Atomists and their forerunners here, because Aristotle is concerned with expounding his own analysis of 'the things that are.' But the implicit criticism is not hard to see. As a program for understanding the world of nature, Atomism is fundamentally misdirected. The *nature* of a thing is not to be found in the material elements that compose it, but rather in the form that determines its composition.

We shall return to give a fuller account of Aristotle's view in volume 2. But it is important at this stage to realize the full force of his position on this matter. He does not say merely that it is important for a philosopher of nature to notice and describe the structure into which the material elements fall when they form compounds. He was in fact ready to concede that Empedocles and Democritus, at least, went a little way in the direction of realizing the importance of structure (*Physics* II.2, 194a20); he had in mind Empedocles' recognition that compounds are determined by the proportions of their component elements (*Parts of Animals* II.1, 642a15ff.), and Democritus' use of the position and arrangement of atoms to account for the perceptible properties of compounds. The point that they missed, which is fundamental to his own philosophy of nature, is that the form or structure of a natural object is not a derivative of the materials that compose it, but a *primary* and *irreducible* factor, not only in the object's being when it exists, but also in its coming-to-be when it is growing.

'Man begets man' is a repeated slogan of Aristotle's. He meant, among other things, to draw attention to the inadequacy of the materialists' program. There is something basically wrong with a program that starts with (say) earth, water, air, and fire, claims that from these elements come more complex beings like wood, foliage, skin, flesh, and bone, and from these again limbs ('Many heads grew, neckless; arms wandered bare, deprived of shoulders,' as Empedocles wrote[1]), and finally from these autonomous limbs grew organized whole creatures. *The form of the organism must be there from the start.* Aristotle's claim was not merely that the form must be grasped and understood by the philosopher who wants to understand nature, but that this is so because forms are in fact the most powerful causes operative in the natural world.

The point emerges clearly in a lengthy criticism of Empedocles, and it is worth quoting since it expresses a theme that is of fundamental importance in this history:

> But it is much more difficult to give an account of generation according to nature. Things that come into being by nature do so either always or usually in such and such a way. Whatever comes into being contrary to this 'always or usually' does so from spontaneity (τὸ αὐτόματον) or luck (τύχη). Now what is the cause of man coming into being from man either always or usually, and wheat rather than an olive from wheat, or bone whenever things are put together in such and such a way? For nothing comes into being by things coming together just as luck has it, as he [Empedocles] admits, but rather when they come together in a certain proportion (λόγος). So what is the cause of these things? Not fire or earth, of course. But not Love and Strife, either, since one is the cause of aggregation only and the other of segregation. The cause is, in fact, the substance of each thing, not just 'mingling and separation of things mingled' as he says. Luck is the name given to those processes, not proportion: mixture can happen just as luck has it. But the cause of things that are as they are by nature is their being of such and such a kind, and this is the nature of each thing. He says nothing about this. (*De generatione et corruptione* II.6, 333b3–18)

In book II of the *Physics* Aristotle sets out his analysis of four types of 'cause': (1) the materials out of which a thing is made; (2) the form; (3) the primary source of change or ceasing to change; (4) 'that for the sake of which' the thing is done. He adds:

> Now, the causes being four, it is the business of the physicist to know about them all, and if he refers his problems back to all of them, he will assign the 'why' in the way proper to his science. (*Physics* II.7, 198a23–5)

He means, not that the physicist should study all four because in that way the right one is bound to be captured, but that a full account of any natural process or natural being must include all four.

This doctrine is picked up in the first book of the *Metaphysics*, in which Aristotle examines earlier philosophy specifically to show that there had been hazy anticipations of his own analysis, that no one had recognized all four of his causes, and that no one had found a fifth.

[1] See above, chapter 7, p. 95.

All of the early philosophers recognized the material cause: some of them recognized only that (1.8, 988b28–31). Presumably Aristotle thought, correctly, that in the Milesians' cosmogony the natural world was produced just by the nature of the primary material elements – moisture for Thales, the Boundless for Anaximander, air for Anaximenes. But some of the later materialists added separate motive causes (1.8, 988b31–3): Anaxagoras' Mind and Empedocles' Love and Strife are his examples. The formal cause was expressed clearly by no one before Plato (1.8, 988b34). The fourth kind of cause, 'that for the sake of which' things are and come-to-be as they are, was in a sense recognized by those who speak of Mind or Love as causes, because they 'class these causes as goods' (1.7, 988b8). But although they thus identified items that might function as final causes, they used them only as starters of motion – i.e. as causes of the second type.

The criticism of the materialists for failing to identify the final cause (the cause that operates as the end – Latin *finis* – of a process) approaches very close to being the same as the criticism that they ignored the formal cause. In Aristotle's own philosophy of nature, it is the form that works as the final cause in the structure of a natural object. Thus the final cause of a bird's possession of wings, for example, is flight, and flight is one of the constituents of the substance or form of a bird – part of what makes this object a bird. So, we can again recognize, in this criticism, Aristotle's main complaint against his materialist predecessors: they failed to understand that form is primary, not a derivative of the elements. This is the focal point of the differences between the cosmologies we are studying.

It can be seen from the quotations in this section that Aristotle was deeply committed to the thesis that the characteristics of material elements alone cannot account for the world of nature. He found in his predecessors, and especially in the Atomists, theories that attempted to derive the natural world from more or less simple elements. He confronts these theories with the thesis that when simple material elements move at random, if they produce anything at all, it is what we call a 'fluke.' If some complex state of affairs, exhibiting some kind of recognizable order and structure, emerges just from the motions of material elements, we recognize it as exceptional, surprising, unexpected. But, he argues, these characteristics are just the opposite of what we find in nature. 'Always or usually' are the characteristic adverbs for natural processes, not 'exceptionally and surprisingly.' Hence it must be that the complex forms of nature did not ever emerge newly from a primitive, unformed condition of the world, and are not now produced by undirected interactions of their material components. Natural forms are as old as the world itself, and in any natural process the form is part of the input, not merely the outcome.

What is surprising to the modern ear in Aristotle's thesis is the assumption that the regularities of nature – the 'always or usually' – cannot be ascribed to the motions of material elements. Why did he think it impossible to state laws of motion for simple elements, such that the complex outcomes of various aggregations of elements would be regular and predictable? Was it not the case that his own theory of the element that composes the stars, planets, sun, and moon allowed for the highest possible degree of regularity and predictability? The answer to these questions must await a fuller examination of Aristotle's own philosophy. For the time being we must give only a partial and provisional answer, on the following lines. Unchangeable elements, he thought, were ruled out for all of the world except the heavens by the phenomena of change; we shall return to this in the next section. But the interactions of changeable elements could be predicted only in very general terms; a quick glance at *Meteorologica* book IV will show how far he was from envisaging an exact science of chemistry or biochemistry. Aristotle looked at the vast gap that separated 'man begets man' from such phenomena as the evaporation of water or the making of cheese, and concluded – perhaps forgivably – that it was a gap that in principle could never be bridged.

13.2 *Aristotle's criticisms of Presocratic theories of elements*

The function of the theories of elements that we have looked at was to offer some kind of explanation of the perceptible properties of the physical world, and especially of their changes. It is on this ground that Aristotle criticizes them: he finds them inadequate, sometimes because they are inconsistent in themselves, more often because they fail to account for what he takes to be the observed facts.

In this book I have tried to show that there was a rather sharp division between theories that preceded Parmenides and those that came after him and took note of his argument. Aristotle did not make the division so sharply, although he was well aware of the actual chronology of the earlier philosophers. It is true that he says explicitly that Leucippus' atomic theory was framed in answer to the Eleatic puzzles (*De generatione et corruptione* I.8, 325a25), and he describes the position of Anaxagoras and Empedocles in terms that emphasize the connection with the Eleatic criticism of 'coming-to-be' (*Metaphysics* I.3, 984a8–16). But he does not notice that the emphasis on the permanence of the elements and the denial of coming-to-be, characteristic of the post-Parmenidean pluralists, is not to be found in the theories of the earlier Milesians.

The reason for this is that Aristotle was bent on finding anticipations of his own theories in all of the early philosophers. He himself held a theory that all physical change presupposes a material substratum that persists through the change. Change – even birth and death – is not adequately described by pointing first to the absence of something and then to its presence. There is always a third factor to be identified. Change must be analyzed as the absence of some property *from some substratum*, then its presence in that substratum; or first its presence, then its absence. Something rather similar to the persisting substratum of this theory could be recognized, he thought, in the single primary element of Milesian theory – Thales' water, Anaximander's Boundless, and Anaximenes' air. We may return briefly to the passage of Aristotle's *Metaphysics* already quoted in chapter 3 (p. 19):

> The element (στοιχεῖον) and principle (ἀρχή) of the things that exist, they say, is that from which all things have their being, from which they come into being in the first place, and into which they perish at the end, *the substance of this persisting while it changes its properties.* (*Metaphysics* I.3, 983b8–10)

This description conceals the difference that I have stressed in earlier chapters between the Milesians' *changing* elements and the *unchanging* elements of the post-Parmenideans. Indeed, Aristotle in this passage goes on to list the Milesians – and adds Empedocles to the list without a hint of major change in the theory. He criticizes the Milesians and post-Parmenideans on different grounds, but that is because the former tried to explain everything as coming from a single element, and the latter did not.

Aristotle's major objection to a single element is that it fails to account for a fundamental feature of the cosmos: the fact of natural motions. On *a priori* grounds he argues that there are just two 'simple magnitudes,' namely the straight line and the circular line. Therefore, there are just three 'simple motions': on the circular line, and in either of the two directions along the straight line. Combining this *a priori* analysis with observation, he concludes that the three simple cosmic motions must be in circular lines around the center (the motion of the heavenly bodies), and in straight lines from the center to the circumference or vice versa. Other kinds of motion must be analyzable as compounds of these. But the simple motions must be characteristic of the simple bodies, he argues. Thus we must assume at least three simple bodies. In fact, his own theory demands five: the element of the heavens has a natural circular motion, earth has a natural motion towards the center, fire away from the center, and there must be two elements intermediate between earth and fire. There are complexities in this theory (for example, we need some explanation of why two opposed linear motions are required, but only one circular motion), which must wait for further treatment in volume 2. This sketch will suffice to indicate the source of Aristotle's objection against the material Monists.[2]

This objection applies to the Atomists, as well as to the early Monists. The thesis of Leucippus and Democritus that atoms are all composed of the same 'stuff' meant, in Aristotle's view, that they could not explain why earth falls and fire rises. We shall look at this point in more detail in section 13.4.

A further criticism raised against some of the Monists accuses them of inconsistency. This applies to those who take the single primary element to be a material that changes its character by rarefaction and condensation: Thales' water and Anaximenes' air are the primary examples, but Aristotle may have supposed Anaximander's 'Boundless' was of the same kind.[3] Aristotle's point is that if the element is *intermediate* in density, it cannot also be *primary*. The reasoning behind this point is that the rare and the dense differ in their texture, the rare being that which is composed of finer – i.e. smaller – parts; but *elements* are the ultimate parts into which a thing can be resolved; hence, it should be that which is resolved into the *smallest* parts that is the elementary body. This is not a criticism that would give any trouble to the Milesians, if, as I believe, they conceived of their 'original stuff' (ἀρχή) not as a variable substratum but as the origin of growth.

Aristotle himself was clear about the need for identifying some factors as primary and others as derivative in the study of nature. At the most fundamental level, as we have seen already, he observed that process in nature required at least three constituents: there must be some subject or substratum that can be identified at the beginning of the process and reidentified at the end of it; and there must be two properties or states of the subject such that the two are in opposition to each other and one is succeeded by the other. The two properties involved in a change may be a pair of 'opposites,' each of them having a name of its own (like 'hot' and 'cold'), or it may be that only one has a name and the other is identifiable only as the 'privation' of the first. These three – subject or substratum, and two opposites – are Aristotle's 'elements' at the most abstract level.

But he was also convinced of the need for 'elements' in another sense: it is correct, he said, to distinguish between compound physical bodies and their simple physical components. It just is the case that some things are made of others, and this relationship is not reversible.

[2] See *De caelo* I.2 for Aristotle's theory, and III.5 for his objections to Monism.
[3] See Cherniss, *Aristotle's Criticism*, pp. 13–14.

Flesh and wood and all other similar bodies contain potentially fire and earth, since the latter can be seen to be extracted from the former. But fire does not contain flesh or wood, either potentially or actually; otherwise they could be extracted from it. (*De caelo* III.3, 302a23–5)

From burning wood and meat come flame and ash (the 'earth' in the quotation above); but there is no similar process by which flame or ash can be made to yield up flesh and wood. The difference between this analysis and the theory of substratum and opposites is that the components discovered by this analysis are perceptible, physical bodies, capable of independent existence.

Having recognized the need for simple bodies, Aristotle uses two sets of criteria for identifying them. One is the distinction of natural motions that we have already discussed; the other arises from the analysis of the simple bodies themselves into substratum and opposites. Hot and cold, dry and moist are generally recognized as the primary pairs of opposites: these can be neatly arranged in pairs to provide a list of simple bodies that coincides with that produced by the criterion of natural motions. Thus fire is light (tends to *rise*), hot and dry; earth is heavy, cold and dry; air is light, hot and moist; water is heavy, cold and moist.

It is against the background of this theory that Aristotle criticizes the Pluralists. Anaxagoras is criticized for his failure to realize that such things as flesh and wood are not themselves elementary and simple but are composed of simple bodies.[4] Aristotle also attacks him for making his elements infinite: this offends against the principle of simplicity, which Aristotle claims to take over from the mathematicians, and is also inconsistent with the observed fact that perceptible qualities, which arise from the simple bodies and their aggregations, are themselves finite.[5] The Atomists are criticized, as we have said, for their failure to account for natural motions, and for introducing infinity in the number of atomic shapes.[6] They are also criticized for making the elements atomic: this is an important argument, which we will reserve for a separate section.

But what of Empedocles, who anticipated Aristotle in giving primary status to earth, water, air, and fire? Aristotle criticizes his theory of elements on two grounds. First, his assumption that the four elements are ungenerated and indestructible is plainly inconsistent with the observed facts: it can be seen that fire dies out, and that water changes to 'air' (i.e. steam or vapour). Aristotle's own theory is based on simple bodies that change into each other, both on the level of the cosmic masses, as in the

[4] *De caelo* III.3, 302a20–b9.
[5] *Ibid.* III.4, 302b10–303a3.
[6] *Ibid.* III.4, 303a4–b8.

rain cycle, and in small quantities. Secondly, Empedocles' theory fails to allow for the distinction between generation and alteration. It was Aristotle's belief that the physical facts demanded this distinction:[7] sometimes an identifiable physical substance changes its properties, and this is 'alteration,' but at other times a new substance is generated from a substratum that does not have the same name as the emergent substance, as when 'air' is generated from water, or a plant from a seed. Empedocles' theory reduced both kinds of change to a mere aggregation and separation of unchanging components, and so failed to capture the true facts. This latter criticism, which is as relevant to the Atomists as it is to Empedocles, is a variation on the theme that we discussed in the preceding section, in that it deals with an aspect of the failure to recognize substantial form or essence.[8]

A striking and major difference between Empedocles and the Atomists, as we have seen, was that the latter stripped away all qualitative properties from their primary elements, leaving only shape, size, and weight. There is, perhaps, less criticism than might have been expected from Aristotle directly on this topic. He mentions that Democritus reduced color to a matter of 'turning' – i.e. the position of atoms[9] – that the white and the dark are equated with the rough and the smooth, and flavors are nothing but shapes (of atoms).[10] He complains that all sensible properties, in Atomist theory, are made into objects of *touch*. What criticism there is here comes from the contrast with Aristotle's own analysis. Touch cannot be all, because we see plainly that there are four other senses. It is wrong to confuse objects accessible to more than one sense with the special objects of each sense. Moreover, sensible properties involve contrariety, such as black and white, bitter and sweet, but different shapes do not fall into contrary oppositions and therefore cannot explain them.[11]

Aristotle found more to be said against the Atomists on the subject of qualitative *change (De generatione et corruptione* 1.8, 325b33–326a24). It was the Atomists' view, of course, that the atoms themselves are totally without change. Aristotle finds a way to attack this through the association of a particular shape with heat. Somewhere in his sources on Atomism he found it said that atoms of spherical shape produce heat. He argues that it cannot be only the shape that produces heat, because in that case there would have to be an opposite shape to produce the opposite

[7] *De generatione et corruptione* 1.1, 314b13.
[8] See *ibid.* 1.2, 317a18–27.
[9] *Ibid.* 1.2, 316a1–2.
[10] *De sensu* 4, 442b1.
[11] *Ibid.* 4, 442a30ff. There is a much more detailed and extended criticism by Aristotle's successor Theophrastus, in his *De sensibus* 68–83.

quality, cold. Hence the spherical atom must produce heat *because it is hot*. But in that case it must follow that atoms can be hotter or colder, or harder or softer (why should they vary in a quality such as heat, but not in other qualities?), and then surely a hotter atom, next to a colder one, must change it by heating it, and a hard atom, hitting a softer one, must change it by denting it, and so on.

These inferences, and others of a similar kind, follow from Aristotle's premiss that the shape of atoms, alone, cannot account for perceptible qualities, because the qualities form pairs of opposites and shapes do not. The Atomists could afford to ignore Aristotle on this subject. Their account of perception included the association of atomic shapes with particular qualities, but there was no such direct correlation as Aristotle assumes. As we have seen in chapter 9, they held that perceptible qualities depend not only on atomic shapes, but also on the position and arrangement of atoms in a compound, and on the condition of the perceiver. If qualities fall into a range between two opposites – and not all of them do – they could be accounted for without postulating opposite shapes.

One particularly shrewd criticism of the atomic theory concerns the notion of contact between two atoms. Atomists needed to assume that atoms could collide with each other, and even if the collision were instantaneous, it must be possible for contact to take place. Aristotle now objects (*De generatione et corruptione* 1.8, 326a30ff.) that if the atoms are all of the same stuff, there can be no difference between two atoms in contact and a single large atom. 'Why, when they come into contact, do they not coalesce into one, as drops of water run together when drop touches drop (for the two cases are precisely parallel)?' This is a very hard question for the Atomists to answer. What separates one atom from another is normally an interval of void, but that can hardly be the case when they are in contact. Atoms were said to be unsplittable because of their 'hardness': would Democritus have claimed that two atoms in contact are splittable from each other because they are not 'hard'? This is an unsatisfactory answer, but there is no surviving evidence to settle the question.

There is a notable mixture of empiricism and the *a priori* about Aristotle's criticisms. Preference for a simple mathematical structure helps to determine the allocation of properties to his elements: they are fully characterized by the distribution of just three pairs of opposites, heavy/light, warm/cold, moist/dry. This pattern rules out both the unlimited differentiation of qualities in Anaxagoras' theory of 'all things together,' and the infinite variety of atomic shapes in the theory of Democritus. At the same time, Aristotle claims that earlier theories fail in various ways to match up to our observation of nature – not, of course, to the carefully controlled observation produced in a scientific laboratory, but to the everyday experience that is expressed in ordinary language. There is too much paradox, for Aristotle, in theories that have to explain away the facts that a fire goes out and water evaporates, or the obvious difference between touch and the other four senses. We shall see later that it is highly characteristic of Aristotle's philosophy of nature to aim at theoretical simplicity without discarding the distinctions and equations of ordinary experience and ordinary language.

13.3 *Criticism of atoms and void*

In the last section we have briefly mentioned some of Aristotle's criticisms of the physical elements of the Atomists, along with other theories of elements. In that context, we noticed his claims that they could not account for natural motions, and that they had no adequate explanation of changes in quality from one opposite to another. We must now look at his objections to the thesis that the physical elements are *atoms* – i.e. bodies that are indivisible.

Aristotle's own analysis of the argument used by Democritus to prove the existence of 'indivisible magnitudes' has been examined in chapter 9 (pp. 123–7). As we saw there, it was an argument derived from Eleatic puzzles about divisibility. It proceeds by first assuming the contradictory of the proposition that there are indivisible magnitudes – namely, that magnitudes are divisible through-and-through – and aims to show that on any interpretation of this assumption, it leads to absurdities. The assumption must therefore be false. Aristotle replies to this that there is a way of interpreting 'divisible through-and-through,' overlooked by the Atomists, that does not lead to any absurdity at all. His new theory of divisibility is set out in reply to the Atomists' argument in *De generatione et corruptione* I.2, and more fully in *Physics* V.3 and book VI; we shall return to it in volume 2. The force of the reply to the Atomists is that a magnitude may be divisible through-and-through in the sense that it can be divided at any point contained in it, but never simultaneously at all the points it contains. This leads to no paradoxes, Aristotle claims, and the thesis of 'indivisible magnitudes' is therefore unnecessary.

Aristotle also aims to prove that the thesis is false. At the beginning of book VI of the *Physics* he shows that no continuous magnitude can be composed of indivisibles. This is a damaging criticism if one assumes, as he does, that there is such a thing as a continuous piece of matter, but it could not trouble the Atomists, who were content to regard all bodies, other than the atoms themselves, as discontinuous aggregates of atoms. It might have raised a difficulty for them if they had proposed to regard space itself as atomic; but it seems unlikely that they did. The effectiveness of this point, then, turns on the persuasiveness of Aristotle's objection, discussed in section 1 of this chapter, that the Atomists could not give a proper account of a unified (i.e. continuous) substance.

There was more difficulty for them in Aristotle's argument that no account can be given of the *motion* of a body that has no distinguishable parts. Although it is a disputed matter, I have maintained in chapter 9 that the ancient authorities who described Democritean atoms as 'partless' were right. If that is so, chapter 10 of Aristotle's *Physics* VI is relevant to them. The argument runs like this. If a thing moves (or changes in any way), we must be able to distinguish a beginning place (or state) and an end place (or state): call them respectively AB and BC. In the time of the motion, the moving thing cannot be in AB, because then it would not have started to move, nor in BC, because then it would have arrived. It must be partly in AB and partly in BC. But, for a partless body, this is impossible. Hence no partless body can be in motion 'except in the sense in which a man sitting in a boat is in motion when the boat is travelling' (VI.10, 241a18) – i.e. when its motion is described in terms of the motion of a continuous whole. Neither Aristotle nor his opponents were able to find a language for talking about motion and rest at an instant. So this argument presented a difficulty. It is significant that the post-Aristotelian Atomists dropped the thesis that atoms are partless; I believe they did so because of Aristotle's argument.[12]

In the history of Greek philosophy, Aristotle was an extremist in his refusal to accept the existence of void in the universe. In defense of his own view that the universe is finite and filled to its limits with a continuous quantity of matter, he mounted several attacks on the void. The attacks came from two directions: from an analysis of the concept of place, and from supposed difficulties in the idea of motion through a void.

On the first count, Aristotle begins with the claim that when people speak of a void, they mean a *place* that is empty: that is to say, somewhere where there might be a body but there happens not to be one (*Physics* IV.6, 213a15). But the place of a thing, according to his own account, is nothing but the inner surface of the body that contains it. Thus the place of the water in a jug is the inner surface of the hollow jug; the place of the earth and sea is the inner surface of the surrounding air. So a void place would be a container with nothing in it. But experience shows that when we speak of emptying a container, we are really describing the replacement of

[12] More about this in volume 2. I have written on this topic in *Two Studies*, pp. 111–30.

one filler by another: if we 'empty' the water out of a jug, we really fill the jug with air instead. If no replacement is possible, then either the original filler refuses to leave, as in the case of clepsydra with the top vent blocked (the illustration is supplied by Simplicius), or else the container collapses, as in the case of an empty wineskin.

This empirical argument is reinforced with further arguments about the concept of place. A void place is possible only if place is conceived as an interval between the surfaces of the container, rather than those surfaces themselves. In his usual manner, Aristotle argues first that the assumption of such an interval is unnecessary in order to do the work demanded of a concept of place, and secondly that it involves certain difficulties on its own account. The first argument (IV.4, 211b18–19) is just the empirical one mentioned already. What we have to describe is always the replacement of one body by another in the same container, not the removal of a filler, leaving the place empty, and *then* the arrival of another filler. If there is a place, it is always the place of one body or the other, and so we do not need to think of place in abstraction from all body. The second argument (IV.4, 211b19–29) is much more complicated, and made more obscure by some doubts about the accuracy of the traditional text. It appears to depend on certain difficulties that arise on the theory of place as an interval when you consider either the parts of the contained substance taken as divisible *ad infinitum* or the parts of something moving in a complex of moved containers. Perhaps all we need to say about it in the present context is this: insofar as it depends on infinite divisibility it would not trouble the Atomists, who denied infinite divisibility; insofar as it depends on puzzles about moved containers, Aristotle's theory was no better off.

Aristotle begins another argument with a pun: 'the so-called vacuum will be found to be really vacuous' (*Physics* IV.8, 216a26ff.). If a solid cube is placed in air or water, it displaces a volume of air or water. But if it is placed in a void, it displaces nothing: the volume of void now occupied by the cube is *coincident* with the volume of the cube. But since neither the void nor the volume of the cube, considered by itself, has any properties other than three-dimensional extension, and this is identical in each of the two, this portion of the void is indistinguishable from the volume of the cube. If two such entities can coincide, however – i.e. the volume of the cube and its volume in void space – why should not any number of them coincide?

The point of this strange argument seems to be that the void is simply not needed at all in our cosmology, if we regard it as an empty space waiting to be filled by something. The cube has its own volume, and its own place, namely the interior surface of the body that contains it. A second, characterless extension coincident with its volume contributes nothing to it, and one might add any number of similar coincident extensions to it without making a scrap of difference to its being what it is and where it is.[13]

The main interest of the argument is to show how far Aristotle was from a notion of space. Whether Leucippus and Democritus were any clearer is doubtful, and this may make us hesitate to accept too clear a distinction in the development of the concept of void.[14]

We come now to Aristotle's arguments against the existence of void from the phenomena of motion. He held that it was impossible to give a rational account of the speed of motion of a body moving through the void, and that no explanation of the direction of natural motions could be given if the universe contains void space.

Other things being equal, he argues (*Physics* IV.8), the speed of a moving body varies inversely with the thickness of the medium through which it moves: the thicker the medium, the slower the movement. For a body of a given size and weight moving over a given distance, we can make a scale correlating the time taken to traverse the distance with the thickness of the medium; for any time taken, there is a corresponding thickness of the medium. But suppose now that this body should move this distance through a void: it must correspond to some (very small) value for the thickness of the medium on our proportion scale. This motion through a void, then, turns out to take the same time as motion through a thin medium. Aristotle takes this to be an impossible result, and concludes that the impossibility results from the supposition that the body may move through a void.

For a better solution to these problems, we have to wait until the sixth century A.D., when Philoponus in his commentary on Aristotle's *Physics* presented a different account of motion and removed the paradoxes about the speed of motion through a void.[15]

But Aristotle's case was not dependent on paradoxes about speed: he protested also that the assumption of an infinite void made it impossible to give a coherent account of natural motion. If the void is supposed to be

[13] Simplicius, I think, interpreted the passage somewhat differently (*Physics* 680–2). He thought Aristotle's point was that if the volume of the solid cube is coincident with the void space it occupies, and both are extensions, then there is nothing to prevent any two extensions from coinciding – for instance, the solid cube and the water that we thought it had to displace. This is unlikely to be right, since it makes the void not 'vacuous' but vicious.

[14] This indicates a doubt I feel about the argument of Sedley, *Phronesis* 27 (1982).

[15] Philoponus' theory of motion will be discussed in volume 2.

infinitely extended in all directions, then no account can be given of a then no account can be given of a center on which natural motions can be focused (*Physics* IV.8, 215a8–9). The Atomists could reply that indeed there is no center, and no centripetal or centrifugal motion, except what is produced by collisions. But, as we have seen, they were able to maintain this only at the expense of ruling out a spherical shape for the earth: a heavy price, indeed. The Stoics did better in this respect when they abandoned the notion of a center of the *universe*, and based their theory of motion on the mutual attraction of all matter in the universe.[16]

Other arguments were directed, not at the infinite extent of the void, but at its characterlessness. Aristotle's theory of place as the inner surface of the containing body meant that any account of locomotion or change of place must be able to give a description of two containing surfaces, one being the place of origin and the other the place of destination. But there are no containing surfaces in a void, since there is no way of differentiating one area of the void from another. 'As there is no difference in what is nothing, there is none in the void (for the void seems to be a non-existent and a privation of being)' (*Physics* IV.8, 215a9–11). This is especially damaging to the idea of natural motion, which Aristotle treats as an indisputable fact.

We have been reviewing arguments directed by Aristotle at the concept of void in abstraction from any particular defenses of it put up by the Atomists. We must now add that he collected a set of arguments that he found in use in favor of the void, and attempted to show that they were not cogent.

It was argued that motion could not take place without void, since nothing that is full can receive anything into itself (IV.6, 213b2). Aristotle replies that this is not so: at least one kind of motion, namely change of quality, can take place in a plenum, and moreover things can give way to each other without there being any void intervals in them or between them, as in the movement of liquids (IV.7, 214a6). This becomes a major point of contention between Aristotle and the Epicureans, as we shall see. Secondly, it is claimed that compression of bodies cannot take place unless there is void in them (IV.6, 213b15). Aristotle replies that it can take place, either because something is squeezed out of the compressed body (IV.7, 214a32), or because compression and rarefaction may be treated as changes in the quality of a continuous matter (IV.9, 217a26ff.). It was argued that growth entails void, because food cannot be ingested

[16] This can be inferred from the rather complex polemics in Plutarch, *De communibus notitiis* 44, supported by Cleomedes, *De motu circulari* I.1.5–6. See Hahm, *The Origins of Stoic Cosmology*, pp. 115–19.

into a body that is already full (IV.6, 213b18). Aristotle replies that this too involves a change of quality (IV.7, 214b1), and in any case would prove too much: since growth takes place in the whole of a body, all of it would have to be void if void were necessary at all. Lastly, it is alleged that a jar full of ash will hold as much water as the same jar empty, and this proves the ash is full of void intervals (IV.6, 213b21). This is a silly argument, and Aristotle does no more than observe that it can be handled in the same way as the third.

13.4 *Rival theories of motion, and the position of the earth*

We have already discussed the main points of Aristotle's differences with his predecessors on motion: as we saw, the main props of his position were the problems he drew out of the idea of motion in a void, and the failure of everybody before him to give an adequate account of natural motions. What remains is to draw the threads together, so as to present this most important issue in clearer outline. But we must begin with some further observations about the motion of the heavenly bodies.

It might be expected that astronomical motions would be the subject within this field on which we should hear the most vigorous protests from Aristotle. His own view was that the regular motions of the heavenly bodies in circular orbits around the earth required that they should be made of special material, endowed with natural circular motion. As we have pointed out in earlier chapters, the Presocratic materialists made no such dualistic assumption. They assumed that the heavenly bodies are made of elements such as are found on earth, and they explained the different behavior of these elements on earth and in the heavens by invoking a special explanation for circular motion. In the post-Parmenidean period, the vortex provided the necessary model.

We should expect, then, that Aristotle might attack the validity of the vortex as a model for explaining the motions of the heavenly bodies. In fact he never quite mounts the comprehensive and systematic attack that we might expect of him. Perhaps the most fundamental criticism is contained in the following passage, of which the first lines have already been quoted above (p. 147):

There are some who suppose that spontaneity (τὸ αὐτόματον) is the cause of this heaven and of all the *kosmoi*. From spontaneity, they say, comes the vortex, that is, the motion that sorted out the universe and brought it into its present order. And this itself is quite astonishing. For they agree that animals and plants neither are what they are, nor grow, by chance (τύχη), but that nature, or mind, or something of that kind is the cause – because it is not any chance thing that grows from each seed, but an olive from this kind, a man from that. But the heaven, and the most divine things of those that are observable, come from chance, they say, and there is no such cause of them as there is of animals and plants. (*Physics* II.4, 196a24–35)

This is paradoxical, Aristotle continues, because in the heavens, which they assign to chance, there is in fact nothing that happens by chance, while in the biological sphere, where they deny chance, there are in fact many chance occurrences.

The heavens are 'the most divine things of those that are observable,' in Aristotle's view, because they are ungenerated and indestructible. The truest god is the unobservable and immaterial cause of movement of the heavens; the heavenly bodies are the observable things that come closest to the immortality of a god. The unchanging regularity of their motions is a witness to their divinity.[17] Thus of all things they are the most inappropriate, in his view, to be explained as the outcome of the chance of the vortex. He repeats the theme in an early chapter of the *Metaphysics*, where he reviews the progress made by earlier thinkers toward an understanding of causes. It could not help being realized, he argues, that material causes are not enough:

For it is not likely that fire or earth or any such element should be the reason why things manifest goodness and beauty both in their being and in their coming-to-be, or that those thinkers should have supposed it was; nor again could it be right to entrust so great a matter to spontaneity (τὸ αὐτόματον) and luck (τύχη). (*Metaphysics* I.3, 984b12–15, trans. Ross, adapted)

This very general criticism is reinforced by a hint of an objection in the first chapter of *De caelo* book II. It seems to me possible that when he wrote this chapter he had in mind some rather fuller exposition of his objection in a work that is now lost: his cosmological dialogue *On Philosophy* would be the likeliest place for it. But that cannot be proved. What we have in this chapter is a link between Aristotle's own theory and ancient mythology, with a glance at recent cosmology in passing. The myths spoke of a kind of divinity, deathless but endowed with motion, and moreover limitless motion – perhaps he refers to the Homeric gods, such as Hermes, who were able to travel where and how they wished, without being subject to human limitations. They associated this divinity with heaven, as being the place where there is no death. Aristotle claims that his own argument, in book I, shows that the motion of the heavens is effortless and unconstrained, and this confirms the mythical picture of heaven. But the myth of Atlas, who held up the sky, must be rejected – and

[17] See Aristotle's *On Philosophy*, fr. 12a (Ross), from Sextus, *Math.* IX.23.

along with Atlas, the vortex of Empedocles. Empedocles believed the heavenly bodies to be heavy, and to be kept up on high and prevented from falling by the force of their whirling motion. But no force of constraint, Aristotle implies, could account for the eternal, unchanging regularity of the motion of the stars.

We have two lines of attack, then, based on the idea that earlier theories involve *chance* and *constraint*, both of these in Aristotle's view being incompatible with what we know about the heavens. Apart from these criticisms, we find objections to the view that the heavens are made of fire (*Meteorologica* 1.3), and to Presocratic theories of such things as comets and the Milky Way (*Meteorologica* 1.6 and 8), but nothing more about the motions of the stars, sun, and moon.

In subsequent history, this remained one of the weakest areas in the defense of Atomism against the Aristotelians. Astronomy provided, through the Hellenistic period, an increasingly exact account of the motions of the heavenly bodies. Aristotle's view of them as 'the most divine things of those that are observable' was confirmed with every step taken by astronomy to prove that apparently irregular motions of the planets, sun, and moon could after all be shown to conform to an exact mathematical model. In Aristotle's geometrical cosmos the heavenly bodies were pictured as being carried in circular orbits on concentric spheres, each rotating on its axis. Later the model was complicated by Ptolemy's addition of epicycles – i.e. orbits that were not concentric with the cosmos – but still motion in a perfect circle remained the basic unit of explanation. In opposition to this, the Atomists could only fall back on analogies with such things as whirlwinds and eddies in running water. On their side, they had the advantage that the model of the vortex was able to offer some kind of explanation of the mechanics of heavenly bodies, whereas the Aristotelian spheres, and still more the Ptolemaic system of circles with epicycles, were exceptionally difficult to make into a plausible mechanical system. Aristotle may at one time have flirted with the idea of a purely mechanical explanation of the rotation of the heavenly spheres.[18] But even if this is true – and it is by no means certain – he gave it up finally in favor of something closer to Plato's *living* cosmos. In this final theory, the heavenly spheres were pictured as alive, and moving because of their love of God. In a religious age, this introduction of God into the cosmic picture was no disadvantage at all. But in any case, the whirls and eddies of the Atomists had small chance of victory with the mathematical astronomers.

Concerning the world inside the heavenly spheres, Aristotle's denial of

[18] See Guthrie, *Aristotle: On the Heavens*, pp. xv–xxxvi.

void did not carry conviction for long. His own successor at one remove, Strato of Lampsacus, modified his definition of place so as to allow the theoretical possibility of an empty place, and later Aristotelians showed that motion through a void could be explained without paradox. But this is a subject for a later chapter.

As to the theory of natural motions of the sublunary elements, we may distinguish two important issues: the question of natural lightness, and the more fundamental question of centrifocal dynamics.

Smoke and flame rise through air; air rises through water. The question was whether these facts pointed to absolute differences, or only to relative ones. The Atomists held the latter alternative: all atoms were heavy, but some were heavier than others, and moreover some compounds had relatively little void in them, others relatively much. Aristotle maintained that a relativist theory would not work: some things must be absolutely light. His case is based on a proposition that he took to be guaranteed by observation:

> A large quantity of fire moves upwards faster than a smaller quantity; and similarly a larger quantity of gold or lead moves downwards ⟨faster than a smaller quantity⟩, and so with all heavy bodies. (*De caelo* IV.2, 309b13–15)

His opponents (including, on this occasion, Plato) explained the difference between heavy and light in three ways: the heavier contains (a) more of the same matter than the lighter, (b) less void than the lighter, or (c) a higher proportion of matter to void than the lighter. The first of these (the only one applicable to Plato) entails that a larger quantity of fire is *heavier* than a smaller quantity and therefore must move upwards more slowly; it also entails that it must be possible for a large quantity of air to be heavier than a small quantity of water. These are both contradicted by the alleged observation just quoted (*De caelo* IV.2, 308b4–28). The second entails that a large quantity of any substance is lighter than a small quantity of the same substance, since it must contain more void; and that is ridiculous (IV.2, 309a17–19). The third, invoking the proportion of matter to void, entails that all quantities of the same stuff are equal in weight, since the proportion is constant for any given stuff; and this is contrary to observation (IV.2, 309b8–17).

We can hardly be critical of Aristotle for being wrong about the speed of fall of more and less heavy bodies, since it took nearly two millennia for the right answer to be discovered. But he should have noticed that a combination of his opponents' thesis (a) with thesis (c) is consistent in itself, and defeats his objections to (a) and (c) separately. At all events, his notion of natural lightness was not found generally acceptable in the Hellenistic period. Like his definition of place, it appears to have been rejected by Strato, who was ready to accept a relativistic account of the distinction (frs. 50–1 Wehrli). There is evidence that the Stoics also abandoned the idea of natural lightness, and the Epicureans certainly rejected it.[19]

Aristotle's centrifocal dynamics, on the other hand, had almost everything in its favor, since the alternative theory that all lines of free fall are parallel seemed to entail that the earth itself is more or less flat. By the end of the fourth century B.C. this had become a position that only last-ditch defenders – the Epicureans – could attempt to hold against the overwhelmingly powerful arguments of the astronomers. But we must go back over previous history to understand how matters stood in Aristotle's lifetime.

In chapter 3 I pointed out that the cosmological myths and the earliest philosophers took for granted what we may call a *linear* or *parallel* picture of the motion of falling bodies. Hesiod could speak of a heavy body falling from heaven to earth, and then – continuing its downward motion – from earth to Tartarus below. The Milesian philosophers abandoned the idea of a Tartarus below the earth, but still asked the question why the earth itself does not fall. They offered suggestions as to what supports it underneath.[20] The earth was thought to be a flat disk, so that lines of fall, vertical to the earth's surface, were all parallel. We have evidence that among later philosophers Anaxagoras, Leucippus, and Democritus also believed the earth to be flat, or saucer-shaped.[21]

Aristotle's *De caelo* articulated the new theory that I have labeled 'centrifocal.' Lines of fall are no longer parallel, but meet at the center of the universe. The verticality of lines of fall to the earth's surface is preserved because the earth itself is pictured as spherical. It is important to notice that it would be very difficult, if not impossible, to combine a centrifocal theory of falling bodies with a flat earth. One has only to imagine a flat earth in the middle of Aristotle's spherical cosmos to see the difficulty. One might suppose that Athens, say, lies at the center of the earth's disk, so that heavy bodies falling to the ground at Athens would fall vertically. But in the Persian capital, or at the Pillars of Hercules, far from the center, lines of fall directed towards the center of the cosmos would be at more or less of an angle to the surface, according to how far the center of the cosmos lay below the earth's surface. The theory would have to be patched up with some remarkable *ad hoc* assumptions.

[19] *Stoicorum Veterum Fragmenta* I.99. Lucretius, *De rerum natura* I.1084–113.

[20] Anaximander is no exception, if my unorthodox view of him is right. See above, chapter 3, note 11.

[21] See the index to Diels–Kranz, vol. III, p. 104a, 13–40 for references.

The vortex model of the formation of the cosmos consorts well with a flat earth, but badly with a spherical earth. The difficulty with the latter is not so much with the shape itself as with the centrifocal dynamics associated with a spherical earth. The vortex, a horizontal spin around a vertical axis, seems to offer no hope of producing out of itself a tendency to move bodies towards the central point of a sphere. It is reasonable to think, as the vortex theorists did, of a tendency for heavy objects in a vortex to collect at the center – meaning the center of the horizontal whirl. Then special reasons were given for the raising of the earth above the *bottom* of the whirl. But this is quite different from the natural centrifocal motions of Aristotle's elements, with his view of the earth as being at rest at the natural destination of heavenly bodies. Although such theories have often been attributed to Democritus, I am unable to understand how the vortex could produce a tendency to move from all directions towards the center of a sphere in bodies that had no such tendency before they were caught in the vortex, nor how the present residue of the vortex – namely, the rotation of the starry sky – could be thought to account for the vertical fall of a heavy body on earth.[22]

In chs. 13 and 14 of *De caelo* II, Aristotle gives a summary of earlier ideas of the position and shape of the earth, his criticisms of them, and a defense of his own theory, always bearing in mind the theory of centrifocal natural motions that he set out in chs. 2–4 of the first book.

One argument for a flat earth, he says, was that the horizon seems to cut the rising and setting sun in a straight line (II.13, 294a1–7); he replies that this fails to allow for the vast distance of the sun from the earth (though it is not clear what this has to do with it), nor for the size of the earth's circumference. It was also suggested that the flatness of the earth accounts for its immobility: it is supported on air like a flat lid (II.13, 294a8–b30). He replies that it is not the shape of the earth but its size that would make this argument work, if it could ever work, and so it is consistent with a spherical shape.

But the major criticism of previous theories arises directly from Aristotle's conviction about the natural motions of the elements. Against the vortex theory, specifically that of Empedocles, he objects that it makes all motion into forced motion, although that makes no sense unless it is contrasted with natural motion. He objects that the vortex cannot explain why heavy objects fall to earth and light objects rise from earth, now that the vortex according to the theory has left the earth stationary at the middle and withdrawn to the circumference of the cosmos. He objects

22 Aristotle raised this latter objection in *De caelo* II.13, 295a33–b1.

that the vortex cannot *produce* the difference between heavy and light objects, but presupposes it (II.13, 294b30–295b9). Finally, in an attack on the theory of Plato's *Phaedo* 108eff. (although he mentions Anaximander – misleadingly, as I have argued in chapter 3), he objects to the idea that the earth stays where it is because it has an equal tendency to move in *all* directions: this would apply to any and all bodies situated at the center, even to fire, and it would be compatible with the earth's moving by expanding equally in all directions from the center (II.13, 295b10–296a21). The *Phaedo* theory is attacked, of course, only as an explanation of the earth's immobility, not of its shape: it assumes a spherical earth.

Aristotle uses his theory of centrifocal motion to argue for the sphericity of the earth, first with the help of a rather unusual thought experiment (297a30–b17). It is assumed that the element earth has a natural tendency towards the center of the universe. If we suppose, then, as the materialists did, that the earth was generated from an original state of dispersal, when the dispersed particles of earth travelled to the center they would naturally fall into a spherical shape, because any anomalies would be self-correcting: a lump on the sphere would be heavier than the counterbalancing portions of it, and so it would continue to press towards the center until all was in balance. The picture is like that of water seeking its own level.

He also uses the observation that all heavy bodies fall 'at equal angles' to the earth's surface: that is to say, the angles between the line of fall and all lines on the earth's surface radiating from the point of impact are all equal. He adds that lines of fall are 'not parallel to each other,' and it must be supposed that by this phrase he means to recall that the lines are directed towards the center of the universe.[23] The only shape for the earth's surface consistent with this geometry is a sphere.

Both of these last arguments take the centrifocal theory as a premiss, and therefore are not effective in themselves against the Atomists and other flat-earth partisans. Aristotle follows up with two arguments of wider application. The first is that eclipses of the moon always show a convex boundary (II.14, 297b24–30). This seems to have little force, since it is not clear why a round disc could not throw a similar shadow. The second is much more powerful: if one travels even quite a short distance north or south, the portion of the sky that is visible changes.

[23] Could Aristotle have *observed* that lines of fall are not parallel? This fact was inferred later by Eratosthenes, from the observation that the vertical coincides with the angle of the sun's rays at noon on midsummer's day at a certain latitude in Egypt, but does not coincide at a point further north at the same time. But I am inclined to think that Aristotle takes it as being given *a priori* by his previously announced theory of centrifocal natural motion.

Some stars are seen in Egypt and around Cyprus but not seen in more northerly places; and in the former, stars set which are continuously visible in the north. It is clear from this, not only that the earth is spherical, but that the sphere is not large; otherwise a small change of position would not produce such an immediately obvious effect. So people who imagine that the Pillars of Hercules [*sc.* Gibraltar] are adjacent to the regions of India, and the ocean is thus all one, are apparently not suggesting anything too much beyond belief. As evidence for what they say they remark also that elephants are a species found at the furthest extremes of both regions, because they are adjacent to each other. Mathematicians who try to calculate the circumference say it is four hundred thousand stades [*sc.* about 74,030 km or 46,000 miles]. (*De caelo* II.14, 297b32–298a17)

This excellent passage is said to have inspired Columbus to set sail west from the Pillars of Hercules.[24] It failed to convince the Atomists, although they could muster only limp defenses against it. Their actual answer, if they gave one, is lost. They could do no more than claim that the effect is like that of walking across the floor of a building with a domed ceiling. This could explain why different portions of the sky appear directly overhead as one moves north or south, but it could hardly explain why portions of the northern sky are actually invisible from further south.

The sphericity of the earth was probably first suggested, as we have seen, by Parmenides, early in the fifth century B.C. It was generally rejected by the materialist philosophers, such as Anaxagoras, and Plato's Socrates in the *Phaedo* speaks as though it were a new and unfamiliar doctrine. The case had grown considerably more powerful by Aristotle's time, because of the astronomical argument we have just examined, and it grew still stronger afterwards in the hands of Hellenistic geographers and astronomers, such as Eratosthenes.

Associated with the sphericity of the earth was the centrifocal dynamic theory. Aristotle's notion of absolute lightness as a centrifugal tendency was soon abandoned, but the rest remained in place. Since it involved the notion of a center and a limiting circumference, it strongly reinforced the belief that our world is a closed and unique system – a belief that was crucial to the development of the medieval Christian world picture. We shall see in a later chapter (volume 2) with what success the Epicurean Atomists argued for the rejection of the centrifocal theory and the unique cosmos, and tried to keep alive the theory of the Infinite Universe.

24 See Ross, *Aristotle*, p. 96, n. 3.

Bibliography

Anton, John P., and Kustas, George L. (eds.). *Essays in Ancient Greek Philosophy*. Albany: State University of New York Press, 1971.

Anton, John P., and Preus, Anthony, (eds.). *Essays in Ancient Greek Philosophy*, vol. II. Albany: State University of New York Press, 1983.

Arnim, Hans von. 'Die Weltperioden bei Empedokles,' in *Festschrift Theodor Gomperz dargebracht zum siebzigsten Geburtstage*, Vienna: A. Holder, 1902, pp. 16–27.

Arrighetti, Graziano. *Epicuro: Opere*. 2nd ed. Torino: Giulio Einaudi, 1973.

Arundel, M. R., *see* Wright, Rosemary Arundel.

Asmis, Elizabeth. 'What is Anaximander's Apeiron?' *Journal of the History of Philosophy* 19 (1981), 279–97.

Bailey, Cyril. *The Greek Atomists and Epicurus*. Oxford, 1928; repr. New York: Russell and Russell, 1964.

Baldes, Richard W. 'Democritus on the Nature and Perception of "Black" and "White".' Phronesis 23 (1978), 87–100.

Balme, D. M. 'Greek Science and Mechanism. I. Aristotle on Nature and Chance.' *Classical Quarterly* 33 (1939), 129–38. 'II. The Atomists.' *Classical Quarterly* 35 (1941), 23–8.

'Development of Biology in Aristotle and Theophrastus: Theory of Spontaneous Generation.' *Phronesis* 7 (1962), 91–104.

Aristotle's De Partibus Animalium I and De Generatione Animalium I (with passages from II.1–3). Oxford: The Clarendon Press, 1972.

Barnes, Jonathan. *The Presocratic Philosophers*. 2 vols. London: Routledge and Kegan Paul, 1979.

'Parmenides and the Eleatic One.' *Archiv für Geschichte der Philosophie* 61 (1979), 1–21.

Beckner, Morton. 'Teleology,' in Edwards, *The Encyclopedia of Philosophy*, vol. VIII, pp. 88–91.

Bicknell, Peter J. 'The Seat of the Mind in Democritus.' *Eranos* 66 (1968), 10–23.

Boas, George. 'Macrocosm and Microcosm,' in *Dictionary of the History of Ideas*, New York: Charles Scribner's Sons, 1968, vol. III, pp. 126a–131b.

Bollack, Jean. *Empédocle*, vols. I–III. Paris: Les Éditions de Minuit, 1965–9.

Bollack, Jean, and Wismann, Heinz. *Héraclite ou la Séparation*. Paris: Les Éditions de Minuit, 1972.

Brentlinger, John. 'Incomplete Predicates and the Two-World Theory of the *Phraedo*.' Phronesis 17 (1972), 61–79.

Burkert, Walter. *Lore and Science in Ancient Pythagoreanism*. Cambridge,

Mass.: Harvard University Press, 1972. Trans. E. L. Minar, Jr with revisions from *Weisheit und Wissenschaft: Studien zu Pythagoras, Philolaos und Platon*. Nuremberg: Verlag Hans Carl, 1962.

'Iranisches bei Anaximandros.' *Rheinisches Museum* 106 (1963), 97–134.

'Das Proömium des Parmenides und die Katabasis des Pythagoras.' *Phronesis* 14 (1969), 1–30.

'La Genèse des choses et des mots: le papyrus de Derveni entre Anaxagore et Cratyle.' *Les Études Philosophiques* 25 (1970), 443–55.

'Air-Imprints or Eidola: Democritus' Aetiology of Vision.' *Illinois Classical Studies* 2 (1977), 97–109.

Burnet, John. *Early Greek Philosophy*, 1st ed. 1892. 4th ed. repr. London: Adam and Charles Black, 1945.

Calogero, Guido. *Studi sull'eleatismo*. Roma: Tipografia del Senato, 1932. German trans. by Wolfgang Raible: *Studien über den Eleatismus*. Darmstadt: Wissenschaftliche Buchgesellschaft, 1970.

Storia della logica antica, vol. 1: 'L'età Arcaica.' Bari: Laterza, 1967.

Canfield, John V. (ed.). *Purpose in Nature*. Englewood Cliffs, N. J.: Prentice-Hall, 1966.

Cherniss, Harold F. *Aristotle's Criticism of Presocratic Philosophy*. Baltimore, The Johns Hopkins University Press, 1935; repr. New York: Octagon Books, Inc., 1964.

Plutarch's Moralia, vol. XII: Concerning the Face which appears in the Orb of the Moon. London: William Heinemann, Ltd, and Cambridge, Mass.: Harvard University Press, 1957.

Classen, Carl Joachim. 'Anaximandros,' in Pauly–Wissowa, *Realencyclopaedie der classischen Altertumswissenschaft*, suppl. 12 (1970), cols. 30–69.

Claus, David B. *Toward the Soul: an Inquiry into the Meaning of Psyche before Plato*. Yale Classical Monographs, 2. New Haven/London: Yale University Press, 1981.

Cole, Andrew Thomas. *Democritus and the Sources of Greek Anthropology*. American Philological Association Philological Monographs, 25. Cleveland: Press of Western Reserve University, 1967.

Cornford, F. M. 'Was the Ionian Philosophy Scientific?' *Journal of Hellenic Studies* 62 (1942), 1–7; repr. in Furley and Allen, *Studies in Presocratic Philosophy*, vol. 1, pp. 29–41.

Principium Sapientiae: the Origins of Greek Philosophical Thought. Cambridge University Press, 1952.

Davis, M. 'Socrates' Pre-Socraticism: Some Remarks on the Structure of Plato's *Phaedo*.' *Review of Metaphysics* 33 (1980), 559–77.

De Ley, H. 'Pangenesis versus Panspermia: Democritean Notes on Aristotle's *Generation of Animals*.' *Hermes* 108 (1980), 129–53.

Dickerson, Richard E. 'Chemical Evolution and the Origin of Life.' *Scientific American* (September 1978), 70ff.

Dicks, D. R. 'Solstices, Equinoxes, and the Presocratics.' *Journal of Hellenic Studies* 86 (1966), 26–40.

Early Greek Astronomy to Aristotle. London: Thames and Hudson, 1970.

Diels, Hermann. *Doxographi Graeci*. Berlin: de Gruyter, 1879; repr. 1958.
Diels, Hermann, and Kranz, Walther. (DK). *Die Fragmente der Vorsokratiker*, 6th ed. Berlin: Weidmann, 1951–2.
Dodds, E. R. *The Ancient Concept of Progress and other Essays on Greek Literature and Belief*. Oxford: The Clarendon Press, 1973.
Duhem, Pierre. *Le Système du monde*, part I: 'La Cosmologie hellénique.' Paris: A. Hermann et Fils, 1913.
Edmunds, Lowell. 'Necessity, Chance, and Freedom in the Early Atomists.' *Phoenix* 26 (1972), 342–57.
Edwards, Paul (ed.). *The Encyclopedia of Philosophy*, 8 vols. New York: Macmillan and the Free Press; London: Collier Macmillan, 1967.
Emlyn-Jones, C. J. *The Ionians and Hellenism: a Study of the Cultural Achievement of Early Greek Inhabitants of Asia Minor*. London: Routledge and Kegan Paul, 1980.
Farrington, B. *Greek Science*, part I, London: Penguin Books, 1944; new ed. 1949; part II, 1949; first publication as one vol. 1953; rev. ed. 1961.
Ferber, Rafael. *Zenons Paradoxien der Bewegung und die Struktur von Raum und Zeit*. Zetemata, 76. Munich: C. H. Beck, 1981.
Ferguson, John. 'Two Notes on the Preplatonics.' *Phronesis* 9 (1964), 98–106.
'DINOS.' *Phronesis* 16 (1971), 97–115.
Ferrari, Giovanni. 'Orality and Literacy in the Origin of Philosophy.' *Ancient Philosophy* 4 (1984), 194–205.
Fränkel, Hermann. 'A Thought Pattern in Heraclitus.' *American Journal of Philology* 59 (1938), 309–37. Abbreviated in Mourelatos, *The Pre-Socratics*, pp. 214–28.
Wege und Formen frühgriechischen Denkens. Munich: C. H. Beck, 1955.
'Studies in Parmenides,' in Furley and Allen, *Studies in Presocratic Philosophy*, vol. II, pp. 1–47. From *Wege und Formen frügriechischen Denkens*, rev. version of 'Parmenidesstudien,' *Nachrichten der Göttingen Gesellschaft* (1930), 153–92.
Frankfort, Henri, *et al. Before Philosophy: the Intellectual Adventure of Ancient Man*. Harmondsworth: Penguin Books Ltd, 1949.
Fritz, Kurt von. *Philosophie und sprachliche Ausdruck, bei Demokrit, Platon und Aristotles*. New York: Stechert, 1938; repr. Darmstadt: Wissenschaftliche Buchgesellschaft, 1966.
'Democritus' Theory of Vision,' in *Science, Medicine and History. Essays on the Evolution of Scientific Thought and Medical Practice Written in Honour of Charles Singer*, vol. I, Oxford University Press, 1953, pp. 83–99. German version in his *Grundprobleme*.
Grundprobleme der Geschichte der antiken Wissenschaft. Berlin: de Gruyter, 1971.
Furley, David J. *Two Studies in the Greek Atomists. Study I: Indivisible magnitudes. Study II: Aristotle and Epicurus on voluntary action*. Princeton University Press, 1967.
'Aristotle and the Atomists on Infinity,' in Ingemar Düring (ed.), *Naturphilosophie bei Aristoteles und Theophrast: Verhandlungen des 4. Symposium Ari-*

stotelicum veranstaltet in Göteborg, August 1966, Heidelberg: Lothar Stiehm, 1969, pp. 85–96. To be reprinted in *Cosmic Problems.*
'Notes on Parmenides,' in Lee, Mourelatos, and Rorty, *Exegesis and Argument*, pp. 1–15. To be reprinted in *Cosmic Problems.*
'Anaxagoras in Response to Parmenides,' in Shiner and King-Farlow, *New Essays on Plato and the Pre-Socratics*, pp. 61–86. Repr. in Anton and Preus, *Essays in Ancient Greek Philosophy*, vol. II, pp. 70–92.
'Aristotle and the Atomists on Motion in a Void,' in Peter K. Machamer and Robert J. Turnbull (eds.), *Motion and Time, Space and Matter*, Ohio State University Press, 1976, pp. 83–100. To be reprinted in *Cosmic Problems.*
'Antiphon's Case against Justice,' in G. B. Kerferd (ed.), *The Sophists and their Legacy: Proceedings of the 4th International Colloquium on Ancient Philosophy*, Hermes Einzelschriften, 44, Wiesbaden: Franz Steiner, 1981, pp. 81–91. To be reprinted in *Cosmic Problems.*
'The Greek Theory of the Infinite Universe.' *Journal of the History of Ideas* 42 (1981), 571–85. To be reprinted in *Cosmic Problems.*
Review of Denis O'Brien, *Theories of Weight in the Ancient World*, vol I: 'Democritus: Weight and Size', in *Oxford Studies in Ancient Philosophy* 1 (1983), 193–209. To be reprinted in *Cosmic Problems.*
Cosmic Problems. Cambridge University Press, forthcoming.
'Truth as What is Unrefuted,' in *Cosmic Problems.*
Furley, David J., and Allen, R. E. (eds.). *Studies in Presocratic Philosophy*, vol. I: 'The Beginnings of Philosophy'; vol. II: 'The Eleatics and Pluralists.' London: Routledge and Kegan Paul, 1970–5.
Furth, Montgomery. 'Elements of Eleatic Ontology.' *Journal of the History of Philosophy* 6 (1968), 111–32; repr. in Mourelatos, *The Pre-Socratics*, pp. 241–70.
Gallop, David. *Parmenides of Elea, Fragments: a Text and Translation with an Introduction. Phoenix* suppl. vol. 18. Toronto/Buffalo/London: University of Toronto Press, 1984.
Gatzemeier, Matthias. *Die Naturphilosophie des Straton von Lampsakos: zur Geschichte des Problems der Bewegung in Bereich des frühen Peripatos.* Monographien zur Naturphilosophie, 10. Meisenheim am Glan: Anton Hain, 1970.
Guthrie, W. K. C. *Aristotle: On the Heavens.* Loeb Classical Library. London: Heinemann, 1939; repr. 1953.
In the Beginning: Some Greek Views on the Origins of Life and the Early State of Man. Ithaca: Cornell University Press, 1957.
A History of Greek Philosophy (*HGP*), vol. I: 'Earlier Presocratics and Pythagoreans'; vol. II: 'The Presocratic Tradition from Parmenides to Democritus.' Cambridge University Press, 1962–5; repr. 1977–80.
Hahm, David E. *The Origins of Stoic Cosmology.* Ohio State University Press, 1977.
Havelock, Eric A. *The Liberal Temper in Greek Politics.* New Haven: Yale University Press, 1957.
Preface to Plato. Cambridge, Mass.: The Belknap Press of Harvard University Press, 1962; repr. 1982.

The Literate Revolution in Greece and its Cultural Consequences. Princeton University Press, 1982.

Heath, P. L. 'Nothing,' in Edwards, *The Encyclopedia of Philosophy*, vol. v, pp. 524–5.

Heath, Sir Thomas. *The Thirteen Books of Euclid's Elements*, 1st ed. 3 vols., Cambridge University Press, 1908; 2nd ed. 1926.

Aristarchus of Samos: the Ancient Copernicus. Oxford, 1913; repr. Oxford: The Clarendon Press, 1959.

Heidel, W. A. 'The *DINE* in Anaximenes and Anaximander.' *Classical Philology* 1 (1906), 279–82.

The Frame of the Ancient Greek Maps, with a Discussion of the Discovery of the Sphericity of the Earth. American Geographical Society Research Series, 20. New York: American Geographical Society, 1937.

'The Pythagoreans and Greek Mathematics.' *American Journal of Philology* 61 (1940), 1–33. Repr. in Furley and Allen, *Studies in Presocratic Philosophy*, vol. I, pp. 350–81.

Heinimann, Felix. *Nomos und Physis: Herkunft und Bedeutung einer Antithese im griechischen Denken des 5. Jahrhunderts.* Schweizerische Beiträge zur Altertumswissenschaft, 1. Basel: Friedrich Reinhardt, 1965.

Hölscher, Uvo. 'Anaximander und die Anfänge der griechischen Philosophie.' *Hermes* 81 (1953), 257–77 and 385–418. Augmented in his *Anfängliches Fragen*, pp. 9–89; abbreviated, in English, in Furley and Allen, *Studies in Presocratic Philosophy*, vol. I, pp. 281–322.

'Weltzeiten und Lebenszyklus: eine Nachprüfung der Empedokles-Doxographie.' *Hermes* 93 (1965), 7–33; repr. with modifications in his *Anfängliches Fragen*, pp. 173–212.

Anfängliches Fragen: Studien zur frühen griechischen Philosophie. Göttingen: Vandenhoeck und Ruprecht, 1968.

Parmenides: vom Wesen des Seienden. Frankfurt am Main: Suhrkamp, 1969.

Huffmeier, Friedrich. 'Teleologische Weltbetrachtung bei Diogenes von Apollonia?' *Philologus* 107 (1963), 131–8.

Hussey, Edward. *The Presocratics.* London: Duckworth, 1972.

Inwood, Brad. Review of Charles H. Kahn, *The Art and Thought of Heraclitus. Ancient Philosophy* 4 (1984), 227–33.

Jaeger, Werner. *The Theology of the Early Greek Philosophers.* The Gifford Lectures, 1936. Trans. Edward S. Robinson. Oxford: The Clarendon Press, 1947; repr. 1948.

Jones, W. H. S. *Hippocrates, with an English Translation.* Loeb Classical Library. Cambridge Mass.: Harvard University Press. London: Heinemann, 1923–31.

Kahn, Charles H. *Anaximander and the Origins of Greek Cosmology.* New York: Columbia University Press, 1960; repr. Philadelphia: Centrum Press, 1985.

'The Thesis of Parmenides.' *Review of Metaphysics* 22 (1969), 700–24.

'On Early Greek Astronomy.' *Journal of Hellenic Studies* 90 (1970), 99–116.

The Verb 'Be' in Ancient Greek. Foundations of Language Supplement Series, 16. Dordrecht: Reidel, 1973.

'Why Existence did not Emerge as a Distinct Concept in Greek Philosophy.' *Archiv für Geschichte der Philosophie* 58 (1976), 323–34.

The Art and Thought of Heraclitus. Cambridge University Press, 1979.

Kerschensteiner, Jula. 'Zu Leukippos A 1.' *Hermes* 87 (1959), 441–8.

Kosmos: Quellenkritische Untersuchungen zu den Vorsokratikern. Zetemata, 30. Munich: C. H. Beck, 1962.

Kirk, G. S. 'Natural Change in Heraclitus.' *Mind* 60 (1951), 35–42. Repr. in Mourelatos, *The Pre-Socratics*, pp. 189–96.

Heraclitus: the Cosmic Fragments. Cambridge University Press, 1954; repr. 1970.

'Some Problems in Anaximander.' *Classical Quarterly* 5 (1955), 21–38; repr. in Furley and Allen, *Studies in Presocratic Philosophy*, vol. 1, pp. 323–49.

Kirk, G. S., and Raven, J. E. *The Presocratic Philosophers: a Critical History with a Selection of Texts*. Cambridge University Press, 1957. 2nd ed. by G. S. Kirk, J. E. Raven, and Malcolm Schofield, 1983.

Knorr, Wilbur R. 'On the Early History of Axiomatics,' in J. Hintikka (ed.), *Proceedings of the Second Conference of the International Union for the History and Philosophy of Science*, Dordrecht: Reidel, 1980, vol. 1, pp. 145–86.

'The Interaction of Mathematics and Philosophy,' in Norman Kretzmann (ed.), *Infinity and Continuity in Ancient and Medieval Thought*, Ithaca/London: Cornell University Press, 1982, pp. 112–45.

Koyré, Alexander. *From the Closed World to the Infinite Universe*. Baltimore: The Johns Hopkins University Press, 1957; repr. New York: Harper and Brothers, 1958.

Kranz, Walther. *Studien zur antiken Literatur und ihrem Fortwirken*, ed. E. Vogt. Heidelberg: Carl Winter Universitätsverlag, 1967.

Laks, André. *Diogène d'Apollonie: La Dernière Cosmologie présocratique. Édition, traduction, et commentaire des fragments et des témoignages*. Cahiers de Philologie, 9. Presses Universitaires de Lille, 1983.

Lämmli, Franz. *Vom Chaos zum Kosmos: zur Geschichte einer Idee*. Schweizerische Beiträge zur Altertumswissenschaft, 10. Basel: Friedrich Reinhardt, 1962.

Lanza, Diego. *Anassagora: Testimonianze e Frammenti*. Firenze: La Nuova Italia, 1966.

Lee, E. N., Mourelatos, A. P. D., and Rorty, R. M. (eds.). *Exegesis and Argument: Studies in Greek Philosophy Presented to Gregory Vlastos, Phronesis* suppl. vol. 1, Assen: Van Gorcum and Co., 1973.

Lennox, J. 'Teleology, Chance, and Aristotle's Theory of Spontaneous Generation.' *Journal of the History of Philosophy* 20 (1982), 219–38.

Lesher, J. H. 'Parmenides' Critique of Thinking: the *polyderis elenchos* of Fragment 7.' *Oxford Studies in Ancient Philosophy* 2 (1984), 1–30.

Lloyd, G. E. R. *Polarity and Analogy: Two Types of Argumentation in Early Greek Thought*. Cambridge University Press, 1966.

'Alcmaeon and the Early History of Dissection.' *Sudhoffs Archiv* 59 (1975), 113–47.

(ed.). *Hippocratic Writings*. The Pelican Classics. Harmondsworth: Penguin Books, 1978.

Magic, Reason, and Experience: Studies in the Origin and Development of Greek Science. Cambridge University Press, 1979.

Science, Folklore, and Ideology: Studies in the Life Sciences in Ancient Greece. Cambridge University Press, 1983.

Löbl, Rudolf. *Demokrits Atome*. Bonn: Habelt, 1976.

Long, A. A. 'Empedocles' Cosmic Cycle in the Sixties,' in Mourelatos, *The Pre-Socratics*, pp. 397–425.

Review of Giovanni Reale, *Melisso: Testimonianze e frammenti*, in *Gnomon* 48 (1976), 645–50.

Long, A. A., and Sedley, David. *The Hellenistic Philosophers*. 2 vols. Cambridge University Press, 1987.

Lonie, I. M. Hippocrates, *The Nature of the Child*, trans. in G. E. R. Lloyd (ed.), *Hippocratic Writings*.

Luria, Salomo. *Die Infinitesimaltheorie der antiken Atomisten. Quellen und Studien zur Geschichte der Mathematik, Astronomie und Physik*, B.2.2, Berlin (1933), pp. 106–85.

Democritea. Leningrad, 1970.

MacKenzie, Mary Margaret. 'Parmenides' Dilemma.' *Phronesis* 27 (1982), 1–12.

McDiarmid, J. B. 'Phantoms in Democritean Terminology.' *Hermes* 86 (1958), 291–8.

'Theophrastus *De sensibus* 61–62: Democritus' Theory of Weight.' *Classical Philology* 55 (1960), 28–30.

McKim, Richard. 'Democritus against Scepticism: All Sense-Impressions are True,' in *Proceedings of the 1st International Congress on Democritus*, Xanthi, 1984, pp. 281–90.

Mansfeld, J. *The Pseudo-Hippocratic Tract 'peri hebdomadon' ch. 1–11 and Greek Philosophy*. Assen: Van Gorcum and Co., 1971.

'Ambiguity in Empedocles B 17.3–5: a Suggestion.' *Phronesis* 17 (1972), 17–39.

Die Offenbarung des Parmenides und die menschliche Welt. Assen: Van Gorcum, 1984.

Marcovich, M. *Heraclitus: Greek Text with a Short Commentary*. Merida, Venezuela: Los Andes University Press, 1967.

Review of Charles H. Kahn, *The Art and Thought of Heraclitus*. *Gnomon* 54 (1982), 417–36.

Martin, Victor. 'Sur la condamnation des athées par Platon au Xe livre des Lois.' *Studia Philosophica* 11 (1951), 103–54.

Miller, H. W. '*Dynamis* and *Physis* in *Ancient Medicine*.' *Transactions and Proceedings of the American Philological Association* 83 (1952), 184–97.

'*Techne* and Discovery in *Ancient Medicine*', *ibid.* 86 (1955), 51–62.

Minar, Edwin L., Jr. 'Cosmic Periods in the Philosophy of Empedocles.' *Phronesis* 8 (1963), 127–45.

Mondolfo, Rodolfo. *L'infinito nel pensiero dei Greci*. Studi Filosofici, 10. Firenze: Felice le Monnier, 1934.

Moorhouse, A. C. '"Den" in Classical Greek.' *Classical Quarterly* 12 (1962), 235–8.

'A Use of "oudeis" and "medeis".' *Classical Quarterly* 15 (1965), 31–40.

Moraux, Paul. *Aristote: du Ciel.* Paris: Les Belles Lettres, 1965.

Morrow, Glenn R. *Plato's Cretan City: A Historical Interpretation of the Laws.* Princeton University Press, 1960.

Mourelatos, Alexander P. D. *The Route of Parmenides: a Study of Word, Image, and Argument in the Fragments.* New Haven: Yale University Press, 1970.

'Heraclitus, Parmenides, and the Naive Metaphysics of Things,' in Lee, Mourelatos, and Rorty, *Exegesis and Argument*, pp. 16–48.

(ed.). *The Pre-Socratics: a Collection of Critical Essays.* Garden City, New York: Anchor Press/Doubleday, 1974.

'Determinacy and Indeterminacy, Being and Non-being in the Fragments of Parmenides,' in Shiner and King-Farlow, *New Essays on Plato and the Pre-Socratics*, pp. 45–60.

'"Nothing" as "Not-Being." Some Literary Contexts that Bear on Plato,' in G. W. Bowersock, W. Burkert, M. Putnam (eds.), *Arkturos: Hellenic Studies presented to Bernard Knox.* Berlin/New York: de Gruyter, 1979, pp. 319–29.

Müller, Carl Werner. *Gleiches zu Gleichem: ein Prinzip frühgriechischen Denkens.* Klassisch-Philologische Studien, 31. Wiesbaden: Otto Harrassowitz, 1965.

O'Brien, Denis. 'The Relation of Anaxagoras and Empedocles.' *Journal of Hellenic Studies* 88 (1968), 93–113.

Empedocles' Cosmic Cycle: a Reconstruction from the Fragments and Secondary Sources. Cambridge University Press, 1969.

'Heavy and Light in Democritus and Aristotle: Two Conceptions of Change and Identity.' *Journal of the Hellenic Society* 97 (1977), 64–74.

Theories of Weight in the Ancient World, vol. I: 'Democritus: Weight and Size'; vol. II: 'Plato: Weight and Sensation.' Paris: Les Belles Lettres, and Leiden: E. J. Brill, 1981–4.

Onians, R. B. *The Origins of European Thought about the Body, the Soul, the World, Time, and Fate.* Cambridge University Press, 1951. 2nd ed. 1954.

Owen, G. E. L. 'Eleatic Questions.' *Classical Quarterly* 10 (1960), 84–102. Repr. in Furley and Allen, *Studies in Presocratic Philosophy*, vol. II, pp. 48–61.

Popper, Sir Karl. 'Back to the Presocratics.' *Proceedings of the Aristotelian Society* 59 (1958–9), 1–24. Repr. with revisions in *Conjectures and Refutations*, London: Routledge and Kegan Paul, 1963, repr. in Harper and Torchback, 1968; and in Furley and Allen, *Studies in Presocratic Philosophy*, vol. I, pp. 130–53.

Pritchard, James B. (ed.). *Ancient Near Eastern Texts Relating to the Old Testament*, 3rd ed. with suppl. Princeton University Press, 1969.

Raven, J. E. 'The Basis of Anaxagoras' Cosmology.' *Classical Quarterly* 4 (1954), 123–37.

Reale, Giovanni. *Melisso: Testimonianze e frammenti.* Firenze: La Nuova Italia, 1970.

Reinhardt, Karl. *Parmenides und die Geschichte der griechischen Philosophie.* 2nd ed. Frankfurt am Main: Klostermann, 1959.

'The Relation between the Two Parts of Parmenides' Poem,' in Mourelatos, *The Pre-Socratics*, pp. 293–311.

Robb, Kevin (ed.). *Language and Thought in Early Greek Philosophy*. La Salle, Illinois: Monist Library of Philosophy, vol. I, 1983.

Robinson, John Mansley. *An Introduction to Early Greek Philosophy: the Chief Fragments and Ancient Testimony, with Connecting Commentary*. Boston: Houghton Mifflin Co., 1968.

'Anaximander and the Problem of the Earth's Immobility,' in Anton and Kustas, *Essays in Ancient Greek Philosophy*, vol. I, pp. 111–18.

Rosenmeyer, Thomas G. 'Plato and Mass Words.' *Transactions and Proceedings of the American Philological Association* 88 (1957), 88–102.

Ross, W. D. *Aristotle's Metaphysics: a Revised Text with Introduction and Commentary*. Oxford: The Clarendon Press, 1924.

Aristotle, 5th ed. London: Methuen and Co., Ltd, 1949; repr. 1953.

Saunders, Trevor J. *Plato: the Laws*, trans. with an introduction. Harmondsworth: Penguin Books, 1970.

Schofield, Malcolm. *An Essay on Anaxagoras*. Cambridge University Press, 1980.

The Presocratic Philosophers, see under Kirk, G. S., and Raven, J. E.

Sedley, David. 'Two Conceptions of Vacuum.' *Phronesis* 27 (1982), 175–93.

Seide, Reinhard. 'Zum Problem des geometrischen Atomismus bei Demokrit.' *Hermes* 109 (1981), 265–80.

Shiner, Roger A., and King-Farlow, John (eds.). *New Essays on Plato and the Pre-Socratics, Canadian Journal of Philosophy*, suppl. vol. II, 1976.

Snyder, Jane M. 'The *Harmonia* of Bow and Lyre in Heraclitus fr. 51 (DK).' *Phronesis* 29 (1984), 91–5.

Solmsen, Friedrich. 'Love and Strife in Empedocles' Cosmology.' *Phronesis* 10 (1965), 109–48; repr. in Furley and Allen, *Studies in Presocratic Philosophy*, vol. II, pp. 221–64.

'The Tradition about Zeno of Elea Re-Examined.' *Phronesis* 16 (1971), 116–41.

Sorabji, Richard. *Necessity, Cause, and Blame: Perspectives on Aristotle's Theory*. Ithaca: Cornell University Press, 1980.

Time, Creation, and the Continuum: Theories in Antiquity and the Early Middle Ages. London: Duckworth, and Ithaca: Cornell University Press, 1983.

Stokes, Michael C. 'Hesiodic and Milesian Cosmogonies.' *Phronesis* 7 (1962), 1–37, and 8 (1963), 1–34.

One and Many in Presocratic Philosophy. Washington: Center for Hellenic Studies. Cambridge, Mass.: Harvard University Press, 1971.

Stückelberger, Adolf. *Antike Atomphysik*. Munich: 1979.

'Lucretius Reviviscens: von der antiken zur neuzeitlichen Atomphysik.' *Archiv für Kulturgeschichte* 54 (1972), 1–25.

Tarán, Leonardo. *Parmenides: a Text with Translation, Commentary and Critical Essays*. Princeton University Press, 1965.

Taylor, C. C. W. 'Pleasure, Knowledge, and Sensation in Democritus.' *Phronesis* 12 (1967), 6–27.

Thomson, George. *Studies in Ancient Greek Society: the First Philosophers*. London: Lawrence and Wishart, 1955.

Tigner, Steven S. 'Empedocles' Twirled Ladle and the Vortex-Supported Earth.' *Isis* 65 (1974), 433–47.

Van der Ben, N. *The Proem of Empedocles' Peri Physios: Towards a New Edition of all the Fragments*. Amsterdam: B. R. Grüner, 1975.

Vernant, J.-P. *Les Origines de la pensée grecque*. Trans. as *The Origins of Greek Thought*. Ithaca: Cornell University Press, 1982.

Vlastos, Gregory. 'Ethics and Physics in Democritus.' *Philosophical Review* 54 (1945), 578–92, and 55 (1946), 53–64. Repr. in Furley and Allen, *Studies in Presocratic Philosophy*, vol. II, pp. 381–408.

'Equality and Justice in Early Greek Cosmologies.' *Classical Philology* 42 (1947), 156–78. Repr. in Furley and Allen, vol. I, pp. 56–91.

'Theology and Philosophy in Early Greek Thought.' *Philosophical Quarterly* 2 (1952), 97–123. Repr. in Furley and Allen, vol. I, pp. 92–129.

Review of F. M. Cornford, *Principium Sapientiae*, in *Gnomon* 27 (1955), 65–76. Repr. in Furley and Allen, vol. I, pp. 42–55.

'One World or Many in Anaxagoras?' in Furley and Allen, vol. II, pp. 354–60; partial repr. of a review of H. Fränkel, *Wege und Formen frühgriechischen Denkens,* in *Gnomon* 31 (1959), 199–203.

'Zeno of Elea,' in Edwards, *The Encyclopedia of Philosophy*, vol. VIII, pp. 368–79.

Plato's Universe. Seattle: University of Washington Press, 1975.

'Plato's Testimony concerning Zeno of Elea.' *Journal of Hellenic Studies* 95 (1975), 136–62.

West, M. L. 'Anaxagoras and the Meteorite of 467 B.C.' *Journal of the British Astronomical Association* (1960), 368–9.

Hesiod: Theogony. Oxford: The Clarendon Press, 1966; repr. 1971.

Early Greek Philosophy and the Orient. Oxford: The Clarendon Press, 1971.

The Orphic Poems. Oxford: The Clarendon Press, 1983.

Wiggins, David. 'Flux, Fire, and Material Persistence,' in Malcolm Schofield and Martha Craven Nussbaum (eds.), *Language and Logos: Studies in Ancient Greek Philosophy presented to G. E. L. Owen*, Cambridge University Press, 1982, pp. 1–32.

Wolf, Michael. *Fallgesetz und Massbegriff: zwei wissenschaftshistorische Untersuchungen zur Kosmologie des Johannes Philoponus*. Quellen und Studien zur Philosophie, 2. Berlin: de Gruyter, 1971.

Woodbury, Leonard. Unpublished paper on Parmenides, presented to the Society for Ancient Greek Philosophy, Toronto, December 1984.

Woodfield, Andrew. *Teleology*. Cambridge University Press, 1976.

Wright, Rosemary Arundel (as M. R. Arundel). 'Empedocles fr. 35.12–15.' *Classical Review* 12 (1962), 109–11.

Empedocles: the Extant Fragments, ed. with an introduction, commentary, and concordance. New Haven: Yale University Press, 1981.

Index of passages

General index

[*This index is intended to supplement the Table of Contents and Index of Passages*]